D1370535

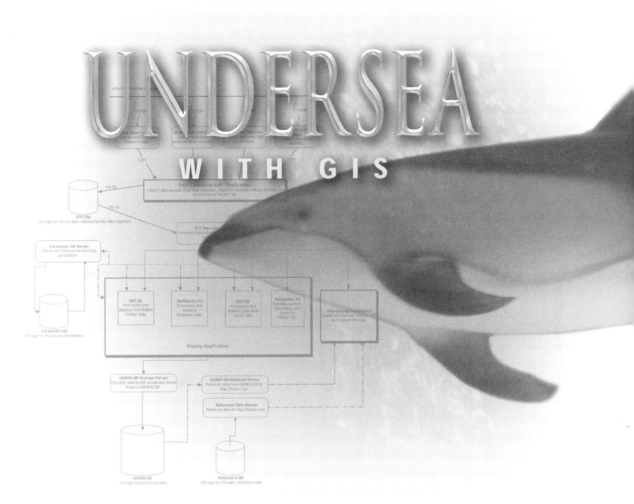

UNDERSEA
WITH GIS

Dawn J. Wright, editor

Foreword by Sylvia Earle

ESRI PRESS

REDLANDS, CALIFORNIA

ESRI
    Undersea with GIS
    ISBN 1-58948-016-3

First printing February 2002.

Printed in the United States of America.

Library of Congress Cataloging-in-Publication Data
Undersea with GIS / Dawn J. Wright, editor ; foreword by Sylvia Earle.
        p.   cm.
    Includes bibliographical references and index.
    ISBN 1-58948-016-3 (pbk.)
    1. Marine sciences-Remote sensing. 2. Geographic information
systems. I. Wright, Dawn J., 1961–
    GC10.4.R4 U54 2002
    551.46'0028'4—dc21   2001008260

Published by ESRI, 380 New York Street, Redlands, California 92373-8100.

Books from ESRI Press are available to resellers worldwide through Independent Publishers Group (IPG). For information on volume discounts, or to place an order, call IPG at 1-800-888-4741 in the United States, or at 312-337-0747 outside the United States.

Contents

Acknowledgments

Gary Amdahl, R. W. Greene, and Christian Harder for excellent editorial assistance; Joe Breman for review of the manuscript; Clem Henriksen for putting the CD so soundly together, and Heather Kennedy for testing it; and Michael Hyatt who designed and Jennifer Johnston who produced the book.

Francesca Cava, Sustainable Seas Expedition (SSE) and National Geographic Society, for encouragement and intellectual stimulation throughout my sabbatical year with SSE, during which time the book was put together.

And finally, to my mother, Jeanne Wright, and my dog, Lydia, for their constant support and inspiration.

Dawn J. Wright, editor

Foreword

SYLVIA EARLE

NATIONAL GEOGRAPHIC EXPLORER-IN-RESIDENCE

"The real voyage of discovery does not consist of seeking new landscapes, but in having new eyes." MARCEL PROUST

Explorers from ages past, from even a few decades ago, would be dazzled by the precision with which we now determine where we are, whether on land, at sea, and even, most amazingly, under the sea. Surely, they would wonder at the extraordinary insights made possible as layers of data are superimposed on precise renderings of terrain, and would readily understand how, in the blink of a geographer's eye, geographic information systems (GIS) have become indispensable to city planners, farmers, businessmen, policy makers, scientists of all sorts—almost anyone who wants or needs to evaluate the scope of human activity, earth's natural processes, or their continuous interplay.

"But what of the ocean," they might ask. "Navigating the surface is one thing, but what's going on below? Is it not possible to provide for an understanding of the sea comparable to what is being done on the land?" In fact, while more has been learned about the nature of the world's oceans in the past twenty-five years than during all preceding human history, most of that vast realm remains unknown, unexplored. While it is known that the sea contains 97 percent of the planet's water, governs climate and weather, generates most of the oxygen in the atmosphere, absorbs carbon dioxide, shapes planetary chemistry, and provides home for most of life on earth, the aquatic two thirds of the planet has been largely neglected.

A new era of ocean exploration is now underway, however, as Dawn Wright makes clear in this remarkable and timely volume. With it, there is a sense of urgency about developing an oceanic GIS, driven largely by the knowledge that the sea, cornerstone of earth's life support system, is being altered significantly by what people are putting into it, and by the huge quantities of wildlife being extracted from it. Concerns are growing about the collapse of once-numerous species of fish and marine mammals, about polluted beaches, toxic algal blooms, "dead zones," and increased occurrence of water-borne diseases. Just as on land, GIS in the sea can facilitate identification of problem areas and point the way toward solutions.

In an effort to safeguard the natural, historic, and cultural heritage of the United States, thirteen marine sanctuaries and a large area of coral reefs in the northwestern Hawaiian Islands have been designated for protection, a counterpart to the nearly four hundred national parks that have been established for the land. Worldwide, more than two thousand marine parks and preserves have been created in response to the growing concerns not just about the declining health of the ocean, but the consequences for humankind. To assess the benefits, determine critical areas, and better manage human activity in the sea, it is vital to know more about the nature of the ocean and how it changes over time.

Classroom globes convey at a glance that the planet is round and mostly blue, but knowledge of the earth as an ever-changing, living system vulnerable to the actions of humankind first came into focus with the advent of high-flying satellites, spacecraft, and exquisite sensors that gather and feed data to increasingly sophisticated computers—and increasingly well-informed human minds.

A growing array of ocean observatories—some tethered as buoys, some drifting, others attached by cables to land stations—is beginning to fill in some of the enormous gaps in our knowledge of ocean temperature, salinity, water chemistry, and biological activity. Acoustic tools make it possible to view and map the ocean floor through sound transmission, reflection, and refraction, while mobile systems—remotely operated vehicles, autonomous unmanned systems, and manned submersibles, as well as thousands of diving human observers—provide vital details.

Some say the most important, influential, and meaningful images to emerge from the twentieth century are those of earth from afar, a brilliant blue and white sphere shining against the dark infinity of space. Perhaps the most important, influential, and meaningful images of the twenty-first century will be derived from millions of data points, a comprehensive view from within earth's blue realm that will help us determine how to find an enduring place for ourselves within the natural systems that sustain us.

Introduction

We have made more progress mapping neighboring planets than we have our own seven seas. We know more about the dark side of the moon and the topography of Venus and Mars than we do of our ocean floors. When NASA officials announced that their high-profile Shuttle Ray Topography Mission had mapped 80 percent of the earth's surface, they neglected to mention that they had skipped the parts that are underwater—71 percent of the globe's actual surface area—which remained impervious to the shuttle's spectacular remote sensors. Recent development, however, of sophisticated technologies for ocean data collection and management has made it possible to imagine—and in some cases to begin to realize—a tremendous potential for mapping and interpreting ocean environments in unprecedented scope and detail. It is specifically with these advancements in mapping, charting, and visualizing in three dimensions both the deep and the shallow ocean that *Undersea with GIS* is concerned.

As agencies and institutes such as the National Oceanic and Atmospheric Administration (NOAA) National Marine Sanctuary Program and National Ocean Service, the U.S. Geological Survey (USGS), and the Monterey Bay Aquarium Research Institute adopt GIS, it's becoming clear that not only are the needs of basic science and exploration being served, but those of ocean protection, preservation, and management as well. Exponential improvements in the speed and capacities of computer hard- and software, an accompanying drop in prices, and the increased availability of skilled practitioners in GIS are making implementation possible where costs have been, until very recently, prohibitive. Data, too, is easier to obtain via the Internet, the World Wide Web, and numerous public sources of spatial information, such as the National Geophysical Data Center, the NASA-funded Distributed Oceanographic Data System (DODS), the EarthExplorer of the USGS, and the Federal Geographic Data Committee's National Geospatial Data Clearinghouse.

And finally, although the realization of true three-dimensionality remains a challenge (particularly in the marine/coastal realm where there are dissimilarities between the horizontal and vertical dimensions), the mapping of our oceans continues to be an area of research that pushes the boundaries of geographic information science, compelling significant attention from funding agencies such as the National Science Foundation and the NOAA Office of High Performance Computing and Communications.

Mapping, charting, and 3-D visualization have been chosen as primary subjects for this book for two reasons. First, the use of GIS for spatial analysis, modeling, and management of oceans has been adequately covered in several recent publications (for example, *Marine and Coastal Geographical Information Systems* from Taylor & Francis, the ESRI Press case-study book *Managing Natural Resources with GIS*, and *Applications of Geographical Information Systems in Oceanography and Fisheries,* forthcoming from Taylor & Francis, as well as conference proceedings such as NOAA's *Coastal GeoTools*). Second, it's still necessary to take a good hard look at the data before proceeding with spatial analyses in GIS, input to numerical models, or preparation of charts for navigation. Mapping and visualization are critical first steps toward identifying patterns in data sets and the underlying processes that created those patterns. They are extremely important as well in assessing the accuracy and utility of one ocean sensor as compared to another, or for interpreting a data set, either singularly or in concert with another.

Undersea with GIS is divided into three sections: part one deals with mapping and visualization, part two with charting, and part three, Internet access. The first section explores several ways in which oceans are currently being mapped in 2-D and "visually experienced" in 3-D. Using the travel time of sound waves (which are transmitted both farther and faster through seawater than the electromagnetic energy used by satellites) to determine depth, and reflectivity (or backscatter) of sound to sense the presence of objects (such as marine mammals or shipwrecks) or to determine hard versus soft spots on the ocean floor, we can dimly "see" in the oceans by "hearing." What we hear, however, is often not clear enough to make sense of the data being collected. We hear it, but we're not sure exactly what it is we're hearing, or what its importance is. Maps made from this kind of data are often aided by tools (today, mainly software) that facilitate visualization and comprehension in the usual two dimensions, but more importantly, spatial relationships and problems in 3-D.

In "ArcView Objects in the Fledermaus Interactive 3-D Visualization System: An Example from the STRATAFORM GIS," Fonseca et al. begin by reviewing the basics of the traditional 2-D map paradigm for GIS, where the two dimensions are horizontal and essentially equivalent. The authors show how this is insufficient for the analysis of complex, 3-D (largely vertical) relationships among data layers in the ocean, then describe the development of an integrated path between the 2-D ArcView® GIS and the Fledermaus interactive, 3-D visualization package. Bathymetric, topographic, and seismic data, water temperature, and many other kinds of data are explored with this integrated system, revealing a number of new geological insights.

"Notes on the Real-Time Interpretation of Seafloor Survey Data" briefly describes how data sets such as those in the previous and other chapters would be obtained during routine oceanographic surveys, reviewing the basic principles of multibeam echo sounding, sidescan sonar surveying, and seismic profiling. A near real-time system for processing, classifying, interpreting, and mapping these kinds of data is then presented. Continuing with the theme of characterizing the ocean floor with GIS, "Seafloor Mapping and GIS Coordination at America's Remotest National Marine Sanctuary (American Samoa)" presents the results of new bathymetric surveys of the Fagatele Bay National Marine Sanctuary in American Samoa, along with a description of efforts to integrate this data with digital video and still photography of the bay's coral reef, which is the main focus of conservation and management within the sanctuary.

Moving on to marine biology, "Using GIS to Track Right Whales and Bluefin Tuna in the Atlantic Ocean" describes efforts at the New England Aquarium to understand linkages between right whale and bluefin tuna distribution and the movements of currents in the Gulf of Maine and the western North Atlantic. GIS analyses in concert with satellite tagging of these species have led to remarkable breakthroughs in the understanding of their behavior and migratory habits.

In "Finding the Green Under the Sea: The Rehoboth Bay *Ulva* Identification Project," Cole et al. describe how Delaware Coastal Programs identified the spatial extent of a problematic species of macroalgae using field surveys, aerial photography, image processing software, and GIS. The resulting maps are aiding resource managers in developing a strategy for containing the macroalgae, and for assessing the broader impact of regional development and agricultural and industrial activities on the aquatic ecology of the bay. In a similar vein, "Geopositioning a Remotely-Operated Vehicle for Marine Species and Habitat Analysis" discusses the management of habitats for marine algae, benthic macroinvertebrates, and finfish within the Punta Gorda Ecological Reserve in northern California. Veisze and Karpov present the development of non-invasive habitat observation techniques using a remotely operated vehicle (ROV), geopositioned with a Trimble differential GPS, positions and video images from which are mapped and hotlinked with ArcView GIS.

Part two looks at recent developments in electronic navigational charts (ENCs, also referred to as electronic nautical charts), which, due to increased benefits in efficiency and safety, may someday completely replace paper charts for the navigation of many vessels. Unlike the other kinds of maps discussed in this book, ENCs are produced only on the authority of organizations sanctioned by the government to do so. They include not only all the spatial, graphical, and text features found on a nautical chart, but additional data from subsequent oceanographic surveys or publications. "The First Three-Dimensional Nautical Chart" briefly reviews the history of ENCs, and the subsequent development of electronic chart display and information systems (ECDIS), from the perspective of the author, Captain Stephen F. Ford, a pioneer in the field. Ford then describes the construction of a prototype for the world's first 3-D nautical chart, made with data from Cape Cod Canal, which would theoretically allow a mariner to actually "see" underwater hazards and obstructions, and thus maneuver his or her vessel in a fashion similar to the operation of an automobile. Recommendations are also made in this regard for improvements in the quality and distribution of oceanographic data from federal agencies, as well as in GIS software.

A system for the management of ENCs, "designed through the collaboration of the Swedish and the Finnish maritime administrations, Novo Meridian Ltd. (Finland), T-Kartor Sweden AB, and ESRI (Redlands, California), is the subject of the next chapter. "HIS: A Hydrographic Information System for the Swedish and Finnish Maritime Administrations," describes the architecture of a hydrographic information system for the management of ENC features, metadata, multiple scales of charts, and data lineage, all with the primary goal of a more efficient ENC production line.

In "Applications of GIS in the Search for the German U-559 Submarine: A Brief Case Study," Cooper et al. do not discuss nautical charts per se, but present issues that would likely be crucial for nautical chart accuracy and production. Nautical charts often include marker symbols for the location of wrecks, and the authors briefly describe the use of a GIS in planning a hypothetical search and salvage operation for a sunken World War II German U-boat in the eastern Mediterranean Sea. Sidescan sonar and bathymetric data were also used to supply information on mobile bottom sediments, possible slumps, and underwater canyons to amplify search parameters and highlight potential false targets.

The third and final section of the book is devoted to ways in which ocean GIS maps and data are being delivered via the Web. In many instances, the central problem with oceanographic data is not its paucity, but the fact that it exists in various formats, scales, and physical locations in a largely inert, non-interactive form (for example, as a graph in a journal publication), unlinked and incompatible with other data sets and models. Because of this, there remains a wealth of observational data, results of experiments, and some data-driven numerical models that have not yet been fully utilized by research scientists, resource managers, and professional educators. The chapters in this section show what progress has been made with Internet map servers that support oceanographic data sharing and provide online spatial analysis tools.

"From Long Ago to Real Time: Collecting and Accessing Oceanographic Data at the Woods Hole Oceanographic Institution" describes three prototype, Web-based applications that were developed to make researchers at the Woods Hole Oceanographic Institution more aware of the capabilities of current GIS technology: an application for geological oceanographers that serves data sets and maps of sediment cores, rocks, and other marine geological artifacts recovered from the seafloor; and two applications for physical oceanographers that allow them to view, query, and download data for the tracking of ocean currents and water mass structure.

Caswell et al. move the discussion from scientific data to underwater cables in "Using the Internet to Manage Geospatial Submarine Cable Data." They discuss a system under development that employs Web-based GIS with C++, Microsoft® Visual Basic®, and several other Web-scripting languages to significantly improve the manner in which geospatial submarine cable data is managed and distributed.

In "Protected Areas GIS: Bringing GIS to the Desktops of the National Estuarine Research Reserves and National Marine Sanctuaries," Killpack et al. discuss the need for a data infrastructure to assist managers in especially sensitive areas of ocean. The Protected Areas GIS Web site includes an Internet GIS mapping application and custom spatial support tools to address specific coastal management issues, such as the optimal siting within a sanctuary for marine reserves (with access that would be even more restricted) and the maintenance of accurate digital boundaries for these regions.

Although a variety of laws, regulations, and special jurisdictions have evolved over time to protect and manage ocean resources, this framework is still extremely vague with regard to the precise positioning of geographic boundaries in the marine environment, where there is really nothing of a fixed, static nature that can be mapped except on the ocean floor. In "Spatial Policy: Georeferencing the Legal and Statutory Framework for Integrated Regional Ocean Management," Treml et al. discuss the implications of current policies, along with the issues surrounding the development of marine boundary GIS data layers in support of regional ocean management and governance. Also described is the Ocean Planning Information System, a Web-based ocean governance and management GIS to facilitate the shift in the United States from fragmented management of individual ocean resources to a more integrated, regionwide approach.

Dragan and Fernetti, in "Geographical Awareness for Modern Travelers: A GIS Application for Maritime Transportation in the Mediterranean Sea," describe the development of the Ship Information and Management System (SIAMS), available either on the Web or via multimedia information kiosks throughout the Mediterranean. SIAMS uses MapObjects® and the MapObjects Internet Map Server to assist international travelers in obtaining real-time ship schedules, retrieving general tourist information on trip destinations, finding connections to other means of transportation, and accessing online booking services (hotels, car rentals, and so forth).

An additional feature of *Undersea with GIS* is the CD–ROM in the back cover, with a variety of free GIS extensions, ARC Macro Language (AML™) scripts, data sets, 3-D flythroughs, and even a sample educational module for K–12 teachers. Many of these tools are direct demonstrations of the concepts and applications described in the chapters of the book (for example, the actual 3-D flythrough of the Cape Cod Canal described in "The First Three-Dimensional Nautical Chart"). They are provided courtesy of ocean GIS specialists from the United States and Great Britain, and it is hoped that they will be suitable for selected research, management, and educational activities.

FAIR WINDS FOR THE FUTURE

Much research remains to be done to improve and extend the capabilities of GIS for the ocean realm. The impediments to further development are not only technological but also conceptual: the lack of complete understanding about the nature of spatial and temporal data continues to obstruct solutions to its manipulation in digital forms (and this is not restricted to oceanographic data). But the ideas and applications presented in this book show that excellent progress is being made. With recent advances in remote sensing technology, marine scientists are now able to apply high-resolution acoustic and optical imaging techniques that span an incredible range of mapping scales, from kilometers to centimeters. And after many years of focus on terrestrial applications, the commercial GIS sector is paying increasing heed to the needs of marine and coastal GIS users, with many leading vendors entering into research and development collaborations with marine scientists, particularly those in state and federal agencies. Recent advances include new cartographic production systems for ocean maps and nautical charts, specific extensions and interfaces for ocean analysis and resource management, fast and intuitive Internet map servers for delivery of products to users, and the design and development of complex oceanographic databases.

Two recent developments by ESRI, the Spatial Database Engine™ (SDE®) and the object-oriented geodatabase concepts in ArcInfo™ 8, hold great promise for managing and modeling ocean features. SDE is essentially a database gateway that allows for the storage and management of spatial data with nonspatial data in a relational database management system, with GIS and tabular attribute data thus stored together, geocoding processing done on the fly, as well as versioning, direct editing of spatial data, and support for new data types such as raster files. As described more fully in the chapter "HIS: A Hydrographic Information System for the Swedish and Finnish Maritime Administrations," this technology is proving to be useful for managing the very large spatial databases that are often created from regional oceanographic surveys (i.e., databases approaching terabyte size that are challenging the capacities of today's computing systems). As for object-oriented geodatabases, these endow geographic features not only with coordinate position and attributes, but also with natural behaviors, so that any kind of relationship may be defined among features. The mobility of many ocean features (such as currents, schools of fish, and hydrothermal plumes) requires representations that are more object-based, with models that are dynamic in terms of position coordinates and time (Lagrangian), rather than fixed (Eulerian).

Finally, the new geodatabase concepts in ArcInfo 8 allow for the creation of standard data models for specific applications and industries, including data classifications, specific data structures, documentation of physical database design, and provision of test databases. It is hoped that marine and coastal applications will join the group of existing ESRI data model development efforts in conservation, defense, energy facilities, forestry, hydrology, parcels, transportation, and water facilities. The publication of this book may provide an exciting segue into these areas. In the meantime, it is my sincerest hope that you will find it to be useful, informative, and fun.

Dawn J. Wright
Department of Geosciences
Oregon State University
Corvallis, Oregon

Chapter 1

ArcView Objects in the Fledermaus Interactive 3-D Visualization System: An Example from the STRATAFORM GIS

LUCIANO FONSECA AND LARRY MAYER

CENTER FOR COASTAL AND OCEAN MAPPING

UNIVERSITY OF NEW HAMPSHIRE

DURHAM, NEW HAMPSHIRE

MARK PATON

INTERACTIVE VISUALIZATION SYSTEMS

FREDERICTON, NEW BRUNSWICK, CANADA

ABSTRACT Advances in acoustic remote sensing technology and marine positioning and orientation techniques have caused the production of, and demand for, marine geospatial data to increase considerably in the last twenty years. These new technologies, when used in concert with other traditional sources of marine information, generate enormous amounts of data and, consequently, distinct challenges in management, analysis, and interpretation. This was the problem faced during the integration of a complex marine database collected off northern California for the Office of Naval Research-sponsored STATAFORM project. In the course of this study more than forty investigators collected a wide range of data aimed at understanding the processes defining continental margin stratigraphy. These data sets were organized into an ArcView GIS database with more than fifty-eight layers and a number of specially developed extensions.

While ArcView GIS was a good way to begin the integration of this large marine database, the 2-D map paradigm of a traditional GIS proved insufficient for a thorough analysis of the complex relationship of the data layers. An interactive 3-D visualization system, Fledermaus, was therefore incorporated. A software filter that translates files from the ArcView GIS database to a Fledermaus internal format was developed. The STRATAFORM database was explored in this 3-D visualization system, revealing a number of new insights.

INTRODUCTION

As advances in acoustic remote sensing technology and marine positioning and vessel orientation techniques have been made (Mayer et al. 1997), so has demand for geospatial data increased. The data sets now being compiled in response to this demand go far beyond traditional sparsely spaced bathymetric measurements, often including very dense multibeam sonar coverage, acoustic imagery, and the use of devices such as sub-bottom profilers, laser airborne depth sounders, and laser line scanners, for example. With these new tools and sources of data it is now feasible to produce much more realistic depictions of the morphology of the seafloor and, as our interpretative and analytical abilities improve, perhaps even thematic information about seafloor character. This technology provides critically needed information for thematic and cartographic mapping, as those activities are applied to natural resource management and exploitation, fisheries habitat studies, environmental monitoring, underwater engineering, geological exploration, and safety of navigation. These new sources of seafloor data, in concert with more traditional sources of marine information, produce enormous amounts of data, creating perforce tremendous difficulties in the proper and effective management, analysis, and interpretation of that data. These challenges, and the demands associated with them, are analogous to those that faced the terrestrial remote sensing community during the 1970s.

Geographic information systems have become the default solution for the integration of large multivariate data sets in almost all terrestrial spatial applications (Aronoff 1991). The same challenge now exists in the marine environment where GIS, with some minor adaptations, can provide a means of integrating large amounts of information dispersed in a variety of media and formats (Wright and Barlett 2000). A GIS approach for underwater applications will offer a straightforward path for visualizing, interpreting, and analyzing vast amounts of data, and for facilitating numerical and logical queries on the database, thereby providing an intuitive means to depict the complex interrelationships among the various data layers.

THE MAP PARADIGM

By definition, a GIS is a computer tool used to collect, store, retrieve, transform, and display spatial data (Burrough and McDonnell 1998). The term *geographic* refers to a known cartographic position of the data (the latitude and the longitude or projection coordinates), and the term *information* refers to attributes assigned to that cartographic location. Consequently, the elemental organization of the GIS database is highly linked to a two-dimensional (2-D) positioning array, which is suitable for a map representation (Bonham–Carter 1996). Following this rationale, the majority of GISs have historically been constrained to two-dimensional data analysis and deal most efficiently with 2-D data sets, although recent 3-D advances from commercial software vendors hold great promise.

Despite this focus on 2-D data sets, most GISs also handle digital elevation models (DEMs), which are representations of continuous surfaces used for topographic, bathymetric, geophysical, and any other data sets that can be described by an array of numbers. Most GISs can present perspective views of the DEM. This is referred to by some authors as 2½-D, because although it is more than a flat 2-D representation, it lacks the formal implementation of a full 3-D setting.

MARINE APPLICATIONS

Although the GIS environment offers an initial solution to many of the problems of data integration, the marine environment poses some unique problems (Lockwood and Li 1995; Wright and Goodchild 1997). First, a system is needed that can take full advantage of, and draw useful inferences from, the especially dense data sets produced by acoustic surveys. Multibeam sonar in shallow water surveys can easily produce tens of millions of soundings (depth measurements) per hour. This enormous and high-speed flow makes for obvious problems in processing and management, quality assurance, and visualization (Mayer et al. 2000).

Another specific requirement of the marine environment is the need to depict information in the water column. Parameters such as temperature, salinity, and sound velocity are measured as a function of depth in different locations, presenting an example of a data cube in a 3-D space. Some new multibeam sonars can detect fish schools and other acoustic targets in the water column. This is a typical example of a 3-D object that must be visualized dynamically within a 3-D scene. The ongoing development of remotely operated vehicles (ROVs) and autonomous underwater vehicles (AUVs) provides a new source of acoustic, photographic, biological, physical, and oceanographic measurements that must be referenced to a 3-D coordinate frame. These vehicles move freely in the water column, acquiring data in all directions.

Engineering projects such as offshore platform building and the laying of pipeline and submarine communication cables need a precise description of the bathymetry and location of natural and man-made obstacles, as well as a characterization of the seafloor and of the shallow sub-bottom. Engineering decisions will be made based on the complex integration of all the information available in a 3-D model (Paton et al. 1997). This is a good example of a network analysis problem that can be addressed in a GIS. The spatial distribution of seafloor relief, terrain derivatives, and estimates of bottom composition are also considered important physical variables governing the distribution of seafloor fish habitats (Martinez 1991). Spatial modeling of these properties in a 3-D environment is a promising method of identifying suitable bottom habitats. Finally, of particular interest is the real-time 3-D visualization of bathymetric data collected for safety of navigation. Real-time visualization can provide tools for controlling quality and designing 3-D electronic charts that provide an intuitive perspective of the relationship of a vessel to navigation hazards.

TOWARD A 3-D GIS

Given the spatial and temporal variability of marine data, the 2-D map concept of the traditional GIS may no longer be suitable for many marine problems. Three-dimensional visualization, on the other hand, permits the very rapid examination and verification of large data sets. During the visualization process, acquisition problems such as data blunders or system artifacts will become evident. In the same way, complex spatial relationships among the data components will be more easily perceived (Ware 1999). Three-dimensional visualization systems improve our capacity for data mining, as the data is presented in a more intuitive way, revealing often-hidden relationships (Kleiner et al. 2000).

Although 3-D visualization can undoubtedly offer tremendous advantages for data analysis, its implementation comes with a high price. Three-dimensional graphical rendering and display require considerable amounts of processing, and until recently were only available in specialized and expensive graphics workstations (MacCullagh 1995). In recent years, however, the hardware and software required for this processing have improved considerably. With increases in the computing power of desktop computers, combined with advances in graphics technology, 3-D graphics are now quite affordable. In concert with these hardware innovations, a number of 3-D scientific visualization software packages capable of running on a variety of platforms have been developed.

The goal for an underwater data integration system is a software and hardware system that combines capabilities for digital mapping, image processing, and database management on one machine at an affordable cost. A cartographic coordinate system must be the basis for reference, with all information attached to a chosen ellipsoid or other earth model. Visualization should also be interactive and fully quantitative, allowing measurements and queries on the displayed database. It should support 3-D topology for searches and spatial analysis, and be attached to a database management system that deals with both spatial and nonspatial data. In short, what's needed is a 3-D GIS that is capable of integrating different data types at different scales, from different sources, with different formats, and covering different areas. While there have been rapid advances in 3-D visualization packages, many of these systems lack the careful geo-referencing and many of the database interrogative capabilities of a complete GIS. Here we present an approach that tries to take advantage of the best of both worlds, by offering an integrated path between the ArcView GIS and the Fledermaus interactive 3-D visualization package.

THE STRATAFORM DATABASE AND GIS

To demonstrate the approach, a complex marine database collected off northern California during the STRATAFORM project is described. STRATAFORM (STRATA FORmation on the Margins) is a multiyear, multi-investigator program funded by the U.S. Office of Naval Research. Its objective is to understand the geologic processes of the shelf and slope responsible for the formation of the sedimentary record over a continuum of scales (Nittrouer and Kravitz 1996). One field area for this program is the highly sedimented and tectonically active Eel River margin off northern California (Nittrouer 1999). In the course of the five years of study of the Eel River margin, an immense database of marine information (including physical oceanographic time series, multichannel seismic data, physical property data from cores, detailed bathymetry, backscatter amplitudes, bottom photos, and a number of other parameters) has been compiled. Each of these distinct data sets provides insight into at least one component of the complex system responsible for generating the strata of the continental margin. This database is remarkable not only for its volume, but also for its diversity.

In response to the need to understand the interrelationships of these data sets, a means was sought to organize and integrate the data into a form that could be easily searched and analyzed—often by investigators working independently. The approach was to treat each data set as an individual layer or theme and bring the data into a widely available GIS. Given that each layer is fully georeferenced and all geodetic corrections (projections, datum, etc.) are applied, the result is the ability to interactively select, explore, retrieve, and display the data sets in any combination desired.

CHOOSING A GIS FOR DATA INTEGRATION

Once it was decided that the STRATAFORM data should be brought into a GIS, the next step was to select the most appropriate package for the task. There were a number of criteria to consider, including the ability to handle both raster and vector data. Most importantly, it was necessary to create a database that would be available to the rest of the STRATAFORM community (and others), without making ownership of the GIS creation software necessary. After evaluating a number of options, ArcView GIS from ESRI was chosen. It has an associated programming language that allows us to customize features to our needs, it supports a number of computer environments (Microsoft Windows® and UNIX for all versions, Apple® Macintosh® to version 3.2), and it provides a freeware viewer (ArcExplorer) that can be distributed with the database, thereby allowing those who do not own ArcView GIS to explore the data, though with some limitations.

More than fifty-eight data layers were collected from a number of STRATAFORM and non-STRATAFORM sources (Mayer et al. 1999). While some sort of geographic information was the common denominator of all the data sets, many came in different formats and were created with different map projections, geodetic datums, and so forth. In order to integrate all of the data sets, they were resolved and reconciled to a common projection and datum framework. The majority of these layers can be displayed with the conventional ArcView GIS tools (figure 1), but certain data sets required display and analysis capabilities beyond those existing in this GIS. To overcome this limitation, a series of Visual Basic programs was developed, which provided extended functionality. These included: (1) a conductivity-temperature-depth (CTD) graphics program, which displays water column profiles recording information on temperature, salinity, and sound velocity (figure 2); (2) a temperature graphics program, which graphs temperature data acquired by current meters deployed for long periods; (3) a seismic viewer, which provides high-resolution visualization of seismic images (figure 3); and (4) an angular response viewer, which displays the acoustic backscatter as a function of the grazing angle for any selected subset of the survey area.

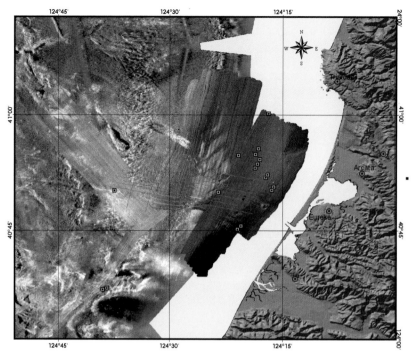

Figure 1. STRATAFORM project in ArcView GIS showing the following layers: (1) Simrad EM1000 multibeam sonar backscatter; (2) EM300 multibeam sonar backscatter; (3) GLORIA sonar backscatter; (4) Vector layer in red showing tracklines from a Huntec high-resolution seismic reflection profiler; and (5) Point data in cyan showing CTD water column profiles recording information on temperature, salinity, and sound velocity.

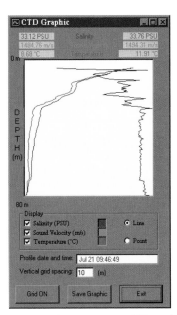

Figure 2. Graph showing CTD profiles. The plotting program is connected to the ArcView GIS layer shown in figure 1 (symbols in cyan). This profile is from the station highlighted in yellow. Each profile can be interactively interrogated.

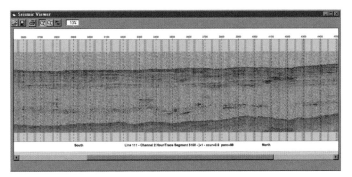

Figure 3. Image showing a Huntec high-resolution seismic profile. This displaying program is connected to the ArcView GIS layer shown in figure 1 (red vectors), and allows zooming and displaying in different scales, taking advantage of the dynamic range of the data. The seismic profile is from the trackline highlighted in yellow in figure 1.

After the database was assembled and functional, the process of data mining and analysis was begun. The need for a fully interactive 3-D environment, able to support all of the multilayer and georeferenced aspects of the GIS, quickly became clear (Van Driel 1995). It was at this point that the Fledermaus interactive 3-D visualization software package from Interactive Visualization Systems, Inc., was chosen.

FLEDERMAUS

Fledermaus is a suite of 3-D visualization tools specifically designed to facilitate the interpretation and analysis of very large (tens to hundreds of megabytes), complex, multicomponent spatial data sets. The goal is to use the 3-D environment to make data available in a natural and intuitive manner that allows the simple integration of multiple components without compromising quantitative aspects.

The software package allows the import of a variety of data types (either gridded or ungridded) from common mapping packages (GRASS, GMT/NetCDF, ISIS), and, with the development of ArcConverter (see the following section), can directly import ArcView GIS layers, as well as ASCII data and a range of binary and image file types. For ungridded data, several approaches to gridding can be chosen from to create a surface representation. Any data brought into the system is fully georeferenced—one can always query the scene to get coordinates and related attributes.

Simple tools allow special maps to be assigned to specific data sets. Already-existing designs can be used, or new ones drawn up on the spot. Once a color map is assigned to the data set, a lighting model is chosen, including artificial sun-illumination, shading, and true shadow calculations. The scene is then rendered to form a 3-D image that is natural looking and easily interpretable, yet fully georeferenced and quantitative. Color, while used to represent depth in the images above, can also be mapped to other parameters (such as backscatter or sediment properties) and draped over the digital terrain model (DTM). Numerous default color maps are provided, as well as the capability to interactively "draw" a color sequence using the hue, saturation, and value (HSV) color space. This allows the user to instantly see how the resulting color schemes enhance, or fail to enhance, the data being viewed.

An interactive shading tool provides control over various lighting parameters such as the direction and angle of ray, apparent glossiness of the surface, and the amount of ambient (background) illumination. A particular feature of the system is the ability to define soft cast shadows. This dramatically increases the ability to perceive certain types of terrain features such as narrow pinnacles and sand waves. The surface data can be shaded directly, or a different data set can be used to map the color onto the surface. In this manner, variables such as multibeam backscatter and sediment type can be instantly draped onto the visualized surface—again, all fully georeferenced.

The user can interactively fly around the data and view it from all angles through the use of a "six degrees of freedom mouse" that transforms simple hand motions into a means of data exploration (and thus the derivation of the name of the software package—Fledermaus). With special LCD glasses, the scene can be viewed in true stereo. Data entry is simple and 3-D interaction can take place within a few minutes of the production of a DTM or sonar mosaic. Thus, if these products are created aboard the vessel, 3-D interaction can be used for quality control, mission planning, and in some cases mission execution.

Through the use of object-oriented software architecture, multiple, individual data sets with different levels of resolution can be easily combined, georeferenced, and displayed. Land-based DEMs with one level of resolution can be combined with multiple bathymetric data, with different levels of resolution, and without the need to regrid at a common scale. Even more intriguing is the ability to texture map data at any level of resolution on top of lower-resolution data. High-resolution backscatter or video imagery can be mapped directly on top of lower-resolution bathymetry without the need to degrade the higher-resolution data or waste memory by upsampling the lower-resolution data. A series of analytical tools is also available that provides for the calculation of gradients, slopes, areas, volumes, and differences, all in the 3-D environment.

Interactive sessions with Fledermaus can be recorded and played back as a movie. With the addition of frame-by-frame video recording equipment, full-resolution videos of exploration sessions can be recorded, or digital-format videos easily created. In addition to the exploration of the data contained in the computer's virtual 3-D environment, Fledermaus also provides a true 3-D stereoscopic display. Normal human stereoscopic vision works only over a limited range. The human eye perceives very little stereoscopic depth at distances greater than 30 meters, and optimal stereoscopic viewing is between 50 centimeters and 2 meters (Ware 1999). Fledermaus incorporates a stereo-viewing algorithm that automatically adjusts the stereo-viewing parameters so that stereoscopic depth is obtained even for scenes of large (virtual) distances. Fledermaus is unique in that these parameters are constantly adjusted so that even while flying through a virtual data environment, stereo depth cues are always available. The primary advantage of a stereo display is that the ability to perceive relative object positions and sizes within the 3-D scene is much greater than in the traditional projection of a 3-D scene onto the computer's 2-D screen.

ARCCONVERTER

ArcConverter is a software filter that transfers data layers from one ArcView GIS project directly into a format that can be interpreted by Fledermaus. It was developed in Visual C++® and is fully integrated in the Fledermaus suite. The first step of the conversion is to parse an ArcView GIS project file (extension .apr) in order to retrieve a list of layers and the location where the files are stored. The retrieved layer list is organized into three basic categories: raster, vector, and grid. The raster category encompasses all images with their geocodification. The vector category includes point, line, and polygon data. The grid category includes all DEMs supported by ArcView GIS.

After the list is retrieved from the project file, one or more layers can be automatically transferred to the Fledermaus 3-D environment. For that, the ArcView GIS files corresponding to the chosen layer must be translated to the tagged data representation (TDR) format of Fledermaus. The TDR format is a framework suitable for storing blocks of information in a file structure. Each TDR file is divided into data blocks, each of which can store a different type of information to be processed. At the beginning of each block there is a tag that uniquely identifies the type and characteristics of the data block (Paton 1995). Thus, a TDR file can basically store any kind of information the user may need, which is extremely convenient in the case of transferring ArcView GIS objects. The following applications will translate ArcView GIS data layers directly into TDR format.

VECTOR DATA

This application will translate ArcView GIS layers with vector data of point, lines, and polygons into the TDR format. ArcView GIS vector data is stored in a shapefile, which is a spatial data format that stores the nontopological geometry and attribute information of spatial features (ESRI 1998). It is actually a set of three different files with the same prefix but different extensions. The geometry of the features is described in the file with the extension .shp. This file has a fixed-size header followed by a variable number of records. Each record describes one geometric entity of variable length by a sequence of vector coordinates. The second file, with an .shx extension, has an index to the feature in the file with extension .shp. The attributes of each spatial feature are stored in a third file with extension .dbf. This file follows the dBASE® format, which is a structured table for database applications. The dBASE table has one record (line on the table) for each corresponding spatial feature in the file with extension .shp. Each record of the table can have any number of fields (column on the table), where each field represents one attribute of the spatial feature. The spatial features and their attributes are transferred to data blocks in a TDR file, where full geometrical and topological structures are preserved. The developed application supports shapefile records of points, polylines (linear vectors), polygons, and null records.

DIGITAL ELEVATION MODELS

This application will parse an ArcView GIS floating-point or integer grid file and then build a DEM, which will be transferred to a TDR file. An ArcView GIS grid file is actually a directory file structure, typically with five or more binary files. The grid data is stored in the first file, called w001001.adf. This file is organized in data tiles with the aim of optimizing access time and storage. The second file, called dblbnd.adf, contains the bounds of the grid in projection coordinates. The header file (hdr.adf) contains the number of tiles into which the grid is divided and the total number of tiles. The file sta.adf contains statistical information about the grid, such as minimum value, maximum value, standard deviation, and so on. The last file (w001001x.adf) is an index to the tiles of the file w001001.adf. Each tile is stored and compressed in a different manner. The actual organization and format of these files and the description of the data tiles is proprietary to ESRI, and any formal documentation concerning these specifications is currently unknown to the authors. The task of unravelling these file formats was partially done by a community of users (Warmendan et al. 2000), from which the information necessary to build the translator was obtained. The application can currently handle the following tile types: constant blocks, 1-bit data, 4-bit data, 8-bit data in sequence, compressed or run-length encoded, 16-bit data in sequence, compressed or run-length encoded, 32-bit data run-length encoded, and floating point.

RASTER IMAGES

This application will read any type of raster image from one ArcView GIS project, with its respective geocodification, and transfer it to a TDR file. In ArcView GIS, images are regarded as a matrix, where only a row and a column are sufficient to reference each cell. In order to proceed with the transformation from image coordinates to projection coordinates, it is necessary to have a four-parameter transformation matrix, since ArcView GIS only handles scale and translation. Any rotation or higher-order distortions have to be done prior to bringing the image into ArcView GIS. This is a major limitation, as it restricts the image to a single cartographic projection. The required geocodification parameters for this transformation are the image bounds in a cartographic projection and the pixel size. The geocodification can be read directly from the image file, which is the case for GeoTIFF, BSQ, BIL, BIP, and ERDAS® formats. In other cases, the geocodification can be read from a second ASCII file, called the world file. This file has the same name as the image file with the letter w added to it. In both cases, the image will be translated to TDR format at the proper location, scale, and projection.

EXPLORING THE 3-D STRATAFORM DATABASE

Illustrated on the next page are both the functionality of the ArcConverter process and the powerful interpretative benefits derived from interactive 3-D visualization by further exploration of the STRATAFORM database. First, an example of a grid transfer is shown. The bathymetric and topographic data sets from the STRATAFORM project came from different surveys and were gridded at different resolutions (figure 4). All of these data sets were ArcView GIS grid files and were converted to Fledermaus TDR files with ArcConverter, thus allowing for full interactive exploration and interpretation (figure 5). Note that the Sea Beam multibeam bathymetric grid has some residual artifacts in the direction of the ship track. These artifacts, related to inappropriate corrections for the sound speed profile, become clear in the 3-D scene.

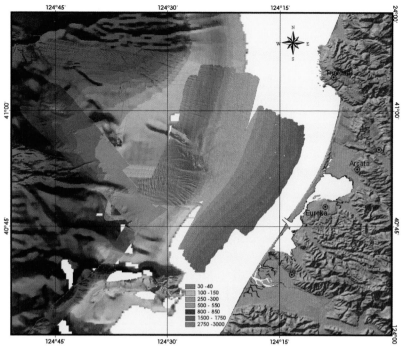

Figure 4. STRATAFORM project in ArcView GIS, showing the following layers:
(1) Sea Beam 12-kHz multibeam bathymetry gridded at 6.48" (~180-meter) resolution;
(2) Atlas Hydrosweep multibeam bathymetry gridded at 2.16" (~60-meter) resolution;
(3) Simrad EM300 30-kHz multibeam bathymetry gridded at a 0.72" (~20-meter) resolution;
(4) Simrad EM1000 95-kHz multibeam bathymetry gridded at a 0.60" resolution; and
(5) a USGS DEM at a 1:100,000 scale.

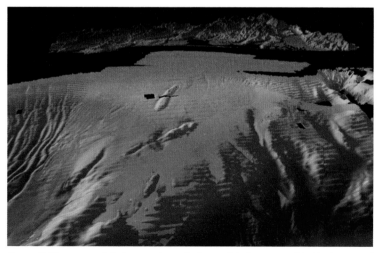

Figure 5. The same data layers shown in figure 4 but converted to Fledermaus format and displayed in Fledermaus.

The raster backscatter mosaics from multibeam and sidescan sonars shown in figure 1 were directly transferred to Fledermaus with ArcConverter. These images are mapped as texture over the bathymetric grids shown in figures 4 and 5. Figure 1 contains vector layers of points (CTD profiles) and of polylines (seismic tracklines), which were also transferred to the 3-D environment. The resulting 3-D view is shown in figure 6. Figure 7 shows a detail from figure 6 where not just its trackline but the actual seismic profile can be seen in the 3-D scene.

Figure 6. The same data layers shown in figure 1 but converted to Fledermaus format.

Figure 7. Detail from figure 6, showing the seismic profile. This line is from the upper portion of the Humboldt slide, showing an undeformed slope stratigraphic sequence and its relation to the regional dip of 3 degrees.

The ability to interactively explore in 3-D the complex relationships between topography, backscatter, and the subsurface seismic data have allowed STRATAFORM scientists to resolve a long-standing debate on the origin of topographic features on the lower continental slope off northern California. The ability to correlate surficial backscatter with subsurface outcrops provided a direct explanation for the origin of both topographic and backscatter targets. Additionally, the ability to directly relate subsurface structure with surficial topography led investigators to conclude that the structures were the result of a massive sediment slide (Gardner et al. 1999).

An example of a polygon layer is shown in figure 8, where Thiessen polygons are drawn around sparse measurements of grain size measured from core samples. A symbol highlights the location of every core. The colors of symbols and polygons are proportional to the grain size. These polygons were converted to Fledermaus format with ArcConverter (figure 9).

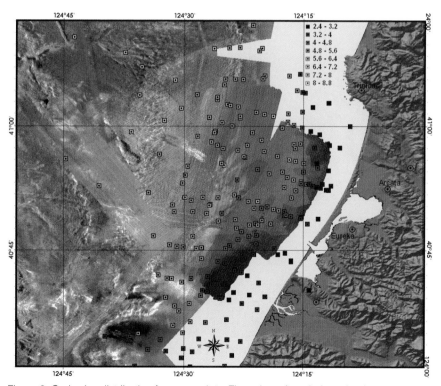

Figure 8. Grain size distribution from core data. The colors of symbols and polygons are proportional to the medium grain size.

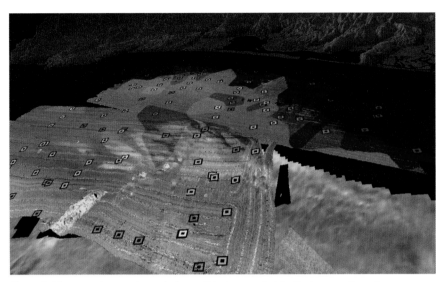

Figure 9. The same layers shown in figure 8 but converted to Fledermaus format.

While 2-D display allows the user to see the relationship between grain size and backscatter (the underlying layer), one cannot see the relationship of both grain size and backscatter to depth. In the 3-D environment one can easily and intuitively resolve the complex relationships between backscatter, grain size, and depth without the need for further analyses.

CONCLUSIONS

The huge amount of information produced by modern acoustic remote sensing systems and other marine surveying tools presents great challenges to management, analysis, and interpretation of marine data. While GISs have become the standard for exploring multiparameter, georeferenced data sets, the 2-D map paradigm of the traditional GIS is often insufficient to allow for the full extraction of the complex interrelationships typical of the marine environment. An approach has been developed that takes full advantage of the well-established capabilities of ArcView GIS, and extends these capabilities through an interface to a powerful, interactive, fully georeferenced 3-D visualization system (Fledermaus). This approach, when applied to a very large and complex marine database collected off northern California, greatly improves our ability to explore and analyze a range of geological and geophysical processes in a relatively simple and intuitive manner.

REFERENCES

Aronoff, S. 1991. *Geographic Information Systems: A Management Perspective.* Ottawa, Canada: WDL Publications.

Bonham–Carter, G. F. 1996. *Geographic Information Systems for Geoscientists: Modelling with GIS.* Computer Methods in the Geosciences Series, Vol. 13. Ontario, Canada: Pergamon Publishers.

Burrough, P. A., and R. A. MacDonnell. 1998. *Principles of Geographical Information Systems: Spatial Information Systems and Geostatistics.* New York: Oxford University Press.

ESRI. 1998. *ESRI Shapefile Technical Description.* ESRI white paper, www.esri.com/library/whitepapers/pdfs/shapefile.pdf, July 1998.

Gardner, J. V., D. B. Prior, M. E. Field. 1999. A larger shear-dominated retrogressive slope failure. *Marine Geology* 154(1–4).

Kleiner, A., L. Gee, and B. Anderson. 2000. Synergistic combination of technologies. *Proceedings of Oceans 2000.* Providence, Rhode Island: Marine Technology Society.

Lockwood, M., and R. Li. 1995. Marine geographic information systems: What sets them apart? *Marine Geodesy* 18(3):157–59.

MacCullagh, J. M. 1995. Power to the people! PC and workstation mapping and database systems. In *Digital Geologic and Geographic Information Systems,* Van Driel, J. N., and J. C. Davis, eds. Washington, D.C.: American Geophysical Union.

Martínez, J. V. 1991. *Analysis of Multibeam Sonar Data for the Characterization of Seafloor Habitats.* Master's thesis. University of New Brunswick, Department of Geodesy and Geomatics Engineering.

Mayer, L., S. Dijkstra, J. H. Clarke, M. Paton, and C. Ware. 1997. Interactive tools for the exploration and analysis of multibeam and other seafloor acoustic data. *SACLANTCEN Conference Proceedings CP-45,* Italy, pp. 355–62.

Mayer, L., L. Fonseca, M. Pacheco, S. Galway, J. V. Martinez, and T. Hou. 1999. *The STRATAFORM GIS CD.* U.S. Office of Naval Research distribution.

Nittrouer, C. A., and J. H. Kravitz. 1996. STRATAFORM: A program to study the creation and interpretation of sedimentary strata on continental margins. *Oceanography* 9(3):146–52.

Nittrouer, C. A. 1999. STRATAFORM: Overview of its design and synthesis of its results. *Marine Geology* 154(1–4):3–12.

Paton, M. A. 1995. *An Object Oriented Framework for Interactive 3-D Scientific Visualization.* Master's thesis, University of New Brunswick, New Brunswick, Canada.

Paton, M. A, L. A. Mayer, and C. Ware. 1997. Interactive 3-D tools for pipeline route planning. *Proceedings of the IEEE–Oceans 97,* 2:1,216–22.

Van Driel, J. N. 1995. Three-dimensional display of geological data. In *Digital Geologic and Geographic Information Systems,* Van Driel, J. N., and J. C. Davis, eds. Washington, D.C.: American Geophysical Union.

Ware, C. 1999. *Information Visualization: Perception for Design.* New York: Morgan Kaufmann Publishers.

Warmerdam, F., et al. 2000. *Arc/Info Binary Grid Format,* gdal.velocet.ca/projects/
aigrid/aigrid_format.html.

Wright, D. J., and M. F. Goodchild. 1997. Data from the deep: Implications for the
GIS community. *International Journal of Geographical Information Systems.* 11(6):
523–28.

Wright, D. J., and D. J. Bartlett, eds. 2000. *Marine and Coastal Geographical Information Systems.* London: Taylor & Francis.

ABOUT THE AUTHORS

Luciano Fonseca is a doctoral student working with Professor Larry Mayer in the
Data Visualization Research Lab of the UNH Center for Coastal and Ocean Mapping, with research interests that include 3-D GIS, visualization of large information
structures, 3-D interactive techniques, multiresolution rendering, database issues,
and practical applications of virtual reality.

Larry Mayer, Ph.D., has had a broad-based background in marine geology and geophysics. He graduated magna cum laude with an honors degree in geology from the
University of Rhode Island in 1973 and received a Ph.D. from the Scripps Institution of Oceanography in marine geophysics in 1979. At Scripps his schizophrenic
future was determined as he worked with the Marine Physical Laboratory's Deep-
Tow Geophysical package, but applied this sophisticated acoustics instrument to
problems of the history of climate. Larry went on to a post-doc at the University of
Rhode Island's School of Oceanography, where he worked on the development of
the CHIRP sonar, as well as problems of deep-sea sediment transport and paleocean-
ography of the equatorial Pacific. In 1982, he became an assistant professor in the
Department of Oceanography at Dalhousie University, and in 1991 moved to the
University of New Brunswick as the NSERC Industrial Research Chair in Ocean
Mapping. In January, Larry became the director of the new Center for Coastal and
Ocean Mapping and the codirector of the NOAA Joint Hydrographic Center at the
University of New Hampshire (UNH). Larry has participated in more than forty-five
cruises (over fifty months at sea!) during the last twenty years, and has been chief or
cochief scientist of numerous expeditions, including two legs of the Ocean Drilling
Program. He has served on, or chaired, far too many international panels and com-
mittees and has the requisite large number of publications on a variety of topics in
marine geology and geophysics. He is the recipient of the Keen Medal in Marine
Geology and was recently awarded an Honorary Doctorate from the University of
Stockholm and appointed to the Presidential Panel on Ocean Exploration. As new
director of the UNH Center for Coastal and Ocean Mapping, Larry's research deals
with sonar imaging, remote classification of the seafloor, and applications of visualization techniques to problems of ocean mapping.

Mark Paton is a senior applications programmer at Interactive Visualization Systems in Fredericton, New Brunswick, Canada, where he is focusing on the development of high-end scientific visualization applications such as interactive 3-D
exploration of ocean floor surfaces, submarine cable routes, and complex climate
data sets. However, the software tools and products that he is helping to develop
also have application in a wide range of fields including environmental impact
assessment, mining, geology, and dredge planning.

CONTACT THE AUTHORS

Luciano Fonseca, Graduate Student Researcher
Telephone: (603) 862-0564
lucianof@cisunix.unh.edu

Larry Mayer, Director
Telephone: (603) 862-2615
larry.mayer@unh.edu

Center for Coastal and Ocean Mapping
Joint Hydrographic Center
Chase Ocean Engineering Lab
24 Colovos Road
Durham, NH 03824
Fax: (603) 862-0839
www.ccom.unh.edu

Mark Paton
Interactive Visualization Systems, Inc.
Incutech Building
University of New Brunswick
P.O. Box 69000
Fredericton, New Brunswick, Canada E3B 6C2
Telephone: (506) 454-4487
Fax: (506) 453-4510
mpaton@ivs.unb.ca
www.ivs.unb.ca

Chapter 2

Notes on the Real-Time Interpretation of Seafloor Survey Data

LARS CHRISTIAN LARSEN

DANISH HYDRAULIC INSTITUTE

HORSHOLM, DENMARK

ABSTRACT ▶ This chapter briefly reviews the research and development efforts of the Danish Hydraulic Institute (DHI) and its partners to provide near real-time presentation of data obtained during seafloor (seabed) surveying and resource mapping. The sensors involved in modern seafloor surveying (e.g., multibeam echo sounders, sidescan sonars, and seismic profilers) typically produce such huge amounts of data that interpretation during sampling becomes prohibitively difficult. Although the individual sensor can be confirmed to work, it is impossible to ascertain the quality of the survey as a whole without going through a complete processing of the data. This is often a very time-consuming effort, during which the normally very expensive survey spread will have to stay mobilized in the area in order to resurvey where data proves inadequate.

INTRODUCTION

It is a well-known fact that the speed and capacity of computers are fast-growing parameters. And because computers (central processing units, microchips, signal processors, etc.) are used extensively in electronic equipment, such devices must also be faster and more capacious. This is also true for equipment used to map the seafloor.

Consequently, the amount of data resulting from typical seafloor surveys is increasing as well, from mere kilobytes/hour to several hundreds of megabytes/hour. To assist in the huge task of interpreting such accumulations of data, the Danish Hydraulic Institute Water & Environment division (www.dhigroup.com) collaborated with RESON A/S Underwater Acoustics (www.reson.com), the Geological Survey of Denmark and Greenland (www.geus.dk), and Rohde Nielsen A/S, a major Danish dredging company (www.rohde-nielsen.dk) on a research project aimed at automating the collection and interpretation of seafloor data. GIS played an important role as well. The results of this project are the basis for this chapter.

SURVEY INSTRUMENTS

The most common reason to commence a survey of an area of the seafloor is to establish a bathymetry for navigational purposes to avoid grounding of ships, or for scientific investigation and interpretation of bottom topography. Other reasons include the desire to search for resources (minerals, gravel, sand, and so forth), conduct environmental investigations (vegetation coverage, mussel banks), or search for objects (wrecks, equipment, pipelines). As direct observation is either impossible or very expensive in most of these cases, seafloors are usually surveyed via various sorts of remote sensing techniques. The equipment we used—a multibeam echo sounder, side scan sonar, and sub-bottom profiler—is designed to provide information on depth, surface texture, and geological condition of the top layers of the seafloor.

MULTIBEAM ECHOSOUNDING

As the name implies, the multibeam echosounder is an echosounder with several beams (figure 1). A basic echosounder measures the depth of water from the surface to the seafloor by sending a pulse of high-frequency sound to the seafloor and measuring the time it takes to be reflected back to the surface. The multibeam echosounder uses the same principle, but uses a large number of receivers with very narrow opening angles, making whole series of measurements possible.

Figure 1. Photographs of typical components in a multibeam echosounding system: (left) transducer head used to emit multiple sound pulses; (right) data collection and processing unit with receiver array and on-screen display of measurements in real time. Reprinted by permission of RESON A/S Underwater Acoustics (www.reson.com).

SIDESCAN SONAR

The sidescan sonar uses the echo sounder principle but is adjusted to transmit the sound pulse sideways. The receiver looks not only at the time lapse of the reflections but also at their strength. By combining this information with a known water depth, a "shadow picture" is generated. The shadow picture will reflect both the texture and the shape of the seafloor (figure 2). For example, hard layers will return strong reflections and large obstacles will show large shadows.

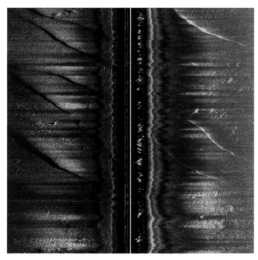

Figure 2. Example of sidescan sonar image showing sandbar and seafloor undulations on both sides of the records, as well as occasional scattered debris.

SUB-BOTTOM PROFILING

In contrast to the sidescan sonar, which uses very high frequencies that will reflect immediately from floor to surface, a sub-bottom profiler uses very low frequencies of sound. These frequencies will penetrate the top layers of the seafloor and reflect from the boundaries between different sediment types. This information can then be interpreted with reference information from cores (borings) and other sources to estimate the types and thickness of individual sediment layers (figure 3).

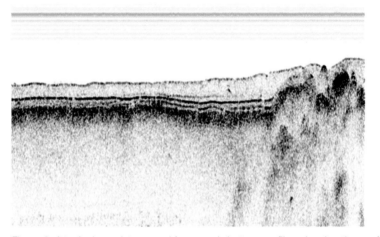

Figure 3. A typical raw data record from a sub-bottom profiler, showing the seafloor and individual sediment layers beneath the seafloor.

SURVEY METHODS

Mapping the seafloor using one or more of the above-mentioned instruments is commonly referred to as a survey. A survey is usually conducted from a single ship with a given set of instruments; the incoming data is logged and treated as a unit to produce the required maps.

NAVIGATION

As mapping is the main objective, every data sample collected needs to be positioned accurately within a coordinate system. As with onshore mapping, surveying at sea is now dominated by the use of the Global Positioning System (GPS), with the same techniques used for onshore mapping being applied to the positioning of the ship.

GPS signals, however, do not penetrate water, making determination of the actual spot on the seafloor that has been measured a more complicated procedure. To accurately position a point (e.g., a spot depth), a system needs to calculate an exact three-dimensional distance from the ship's GPS antenna to that particular point on the seafloor.

A detailed explanation of the difficulties and the numerous methods attempted in this context is beyond the scope of this chapter (rather see works such as Blondel and Murton 1997; Wright 1999). In general, suffice it to say that no system is completely accurate, especially in the case of instruments towed behind the ship, where another factor of uncertainty becomes involved. Therefore, in addition to interpretation of data received from the scanning instruments, a detailed evaluation of the positioning of that data must also take place.

INTERPRETATION OF DATA

Just a few years ago the output from scanning instruments would be in the form of paper records with images of the return signal. An expert would then manually annotate these records, highlighting the features to be mapped. Today the standard is to record the information digitally, but as each instrument produces several hundred megabytes per day, even the fastest computers struggle to manage such voluminous flows.

During the period 1998–2000, DHI Water & Environment, working with the industry and research partners mentioned above, carried out a project in which data would not only be handled in real time but interpreted as well. To do so would mean much faster turnaround times for surveys, as well as guaranteeing that the survey ship and instruments not be docked until all the data collected has been confirmed as usable.

Real-time interpretation of sidescan data is especially challenging. The data is represented as images showing reflections from the seafloor, but the reflections are due to physical conditions varying not only with the nature of the target but also with distance from the sensor. Because of this—and many other factors as well, including the fact that acoustic beams must be angled—it has proven very difficult to achieve automated interpretation on a global scale.

The method described below and shown in figure 4 is one attempt. It incorporates pattern recognition and digital image processing, along with the application of artificial neural networks in connection with so-called self-organizing maps. First, the sidescan image is divided into a number of smaller rectangles, as indicated in the top middle of figure 4. For each of these subrectangles, a number of characteristic features are calculated and used as the basis for classification. Most of the extracted features concern the texture properties of the seafloor. Texture is reflected in the pictorial pattern produced by the spatial distribution of gray-level variations. The two most important texture properties are tone, which describes the primitives out of which the texture is composed, and surface texture, as reflected in the spatial dependencies between the primitives. Measures for these properties may be extracted from the bitmap by calculating other properties, such as energy, entropy, and momentum. A neural network approach is used to divide a training data set into a finite number of classes; characteristics from the training data set are then used to map real measurements into classes by a relatively simple (and fast) mapping algorithm.

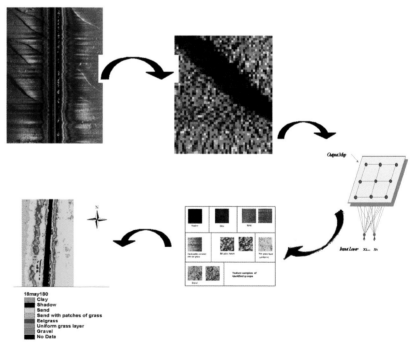

Figure 4. Methodological steps in interpreting sidescan sonar imagery (see text for explanation).

DATA FLOW AND PROCESSING

As mentioned before, there will always be a relatively high uncertainty in the positioning of data coming from surveying instruments. In order to involve information from all sensors in a holistic interpretation, a DHI system called GENIUS creates an interpretive basis by accessing several other reference databases at once. These can be previous survey results from the same area, as well as geological databases containing results from coring operations and other kinds of seafloor sampling projects. The general data flow of the system is shown below in figure 5.

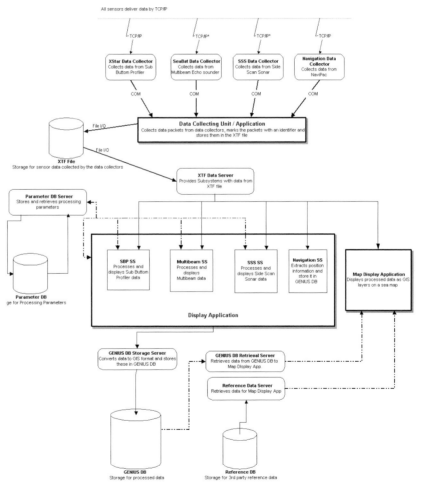

Figure 5. Data flow and processing diagram for DHI's GENIUS system, which incorporates multibeam bathymetry, sidescan sonar, and sub-bottom profiler data and includes GIS map display and analysis capabilities.

Because the output formats of survey instruments differ, the first steps in the preprocessing of data in GENIUS include various formatting and reformatting procedures. The most important task, however, is to synchronize the incoming data with respect to time and position, and to produce a playback file that can be used if the following processing steps fail.

Figure 5 also shows parallel flows of data to two applications. One is the display and editing application, in which both automated and operator-assisted interpretations are made. The other is a general map display (figure 6) where the user can choose various layers of information to match the incoming data with reference data from earlier surveys and other background databases. Both applications make extensive use of GIS applications and tools to handle the data. They are based specifically on ArcView 8 and ESRI's Spatial Database Engine (ArcSDE™).

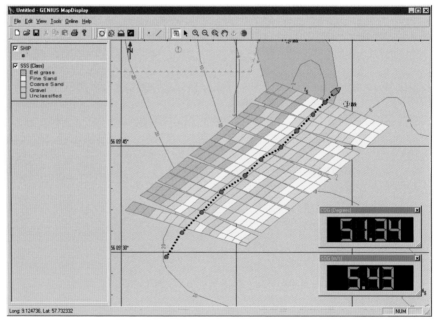

Figure 6. Screen capture of the map display portion, based on ArcView 8, of the GENIUS system. The map display shows the track of a research ship along which regions of the seafloor have been identified and classified according to bottom type (e.g., eel grass, sands, and gravels).

CONCLUSION

A brief review of basic seafloor surveying instruments and techniques has been presented along with a description of how the resulting data is processed, interpreted, and mapped by the GENIUS system of DHI. A primary objective reached by DHI and its partners was the handling, interpretation, and classification of data in near real time, the implications of which include faster turnaround times and the relative certainty that a survey ship and its instruments will not be docked until the data it collected has been confirmed as usable. These advancements in effectiveness of labor and efficiency of process can mean savings not only of time but of money as well—tens of thousands of dollars per project, making surveying at sea much more feasible than at present.

REFERENCES

Blondel, P. and B. J. Murton. 1997. *Handbook of Seafloor Sonar Imagery*. Chichester, UK: John Wiley & Sons.

Wright, D. J. 1999. Getting to the bottom of it: Tools, techniques, and discoveries of deep ocean geography. *The Professional Geographer* 51(3):426–439.

ABOUT THE AUTHOR

Lars Christian Larsen is a chief engineer in the informatics department of DHI Water & Environment, Denmark. He has been working with online monitoring and data collection in the marine environment for more than twenty years. Presently he is serving as manager of GIS development for DHI.

CONTACT THE AUTHOR

Lars Christian Larsen
Danish Hydraulic Institute
Agern Alle 5
Horsholm, DK2970
Denmark
Telephone: +45 4517 9357
Fax: +45 4517 9200
lcl@dhi.dk

Chapter 3

Seafloor Mapping and GIS Coordination at America's Remotest National Marine Sanctuary (American Samoa)

DAWN J. WRIGHT
DEPARTMENT OF GEOSCIENCES
OREGON STATE UNIVERSITY
CORVALLIS, OREGON

BRIAN T. DONAHUE AND DAVID F. NAAR
CENTER FOR COASTAL OCEAN MAPPING
UNIVERSITY OF SOUTH FLORIDA
ST. PETERSBURG, FLORIDA

ABSTRACT ▶ Currently there are thirteen sites in the U.S. National Marine Sanctuary System that protect more than 18,000 square miles of American coastal waters. Coral reefs are a particular concern at several of these sites, as reefs are now recognized as constituting some of the most diverse and valuable ecosystems on earth, as well as being the most endangered. The smallest, remotest, and least-explored site is the Fagatele Bay National Marine Sanctuary (FBNMS) in American Samoa, the only true tropical coral reef in the sanctuary system. It was largely unexplored below depths of 30 meters, and no comprehensive documentation of its plants, animals, and submarine topography existed. Very little is known generally of shelf-edge coral reef habitats (approximately 50–100 meters deep) throughout the world, and no inventory of benthic-associated species has been compiled. This chapter presents the results of (1) recent multibeam bathymetric surveys in April and May, 2001, which were conducted to obtain complete topographic coverage of the deepest parts of FBNMS, as well as other sites around the island of Tutuila, and (2) efforts to integrate this and other baseline data into a GIS to facilitate future management decisions and research directions within the sanctuary.

INTRODUCTION   In 1972, in the face of rising coastal development, increasing pollution, and accelerated species extinction rates, the National Marine Sanctuary System was created to protect ecological, historical, and aesthetic resources within vital areas of U.S. coasts (www. sanctuaries.nos.noaa.gov). Currently there are thirteen official sanctuaries protecting more than 18,000 square miles of American coastal waters from American Samoa to Maine and the Florida Keys, including Pacific and Atlantic habitats for whales, sea lions, rays, turtles, kelp forests, and coral reefs (Earle and Henry 1999). Activities at these thirteen sites will be further supported by the Oceans Act of 2000, established in the summer of 2000 by President Clinton to develop and enact recommendations for strengthening and coordinating federal ocean policy (e.g., dusk.geo.orst.edu/undersea/oc2K.pdf). More than thirty years ago, a similar presidential act led to the creation of the National Oceanic and Atmospheric Administration (NOAA).

Coral reefs are a particular concern at several of the sanctuaries as they are now recognized as some of the most diverse and valuable ecosystems on earth. Reef systems are storehouses of immense biological wealth and provide economic and ecosystem services to millions of people as shoreline protection, areas of natural beauty and recreation, and sources of food, pharmaceuticals, jobs, and revenues (Jones et al. 1999; Wolanski 2001). Unfortunately, coral reefs are also among the most threatened marine ecosystems on the planet, having been seriously degraded by human overexploitation of resources, destructive fishing practices, coastal development, and runoff from improper land-use practices (Bryant et al. 1998; Wolanski 2001).

A major initiative, administered by NOAA, has recently been launched to explore, document, and provide critical scientific data for the sanctuaries. The goal is to develop a strategy for the restoration and conservation of the nation's marine resources (Bunce et al. 1994; Wilson 1998). One of the major catalysts behind this effort is the five-year Sustainable Seas Expeditions project (SSE, sustainableseas.noaa.gov), led by marine biologist and National Geographic Explorer-in-Residence

Dr. Sylvia Earle and former National Marine Sanctuary Program Director Francesca Cava. SSE has been using the one-person submersible, DeepWorker, to pioneer the first explorations of the sanctuaries. Its mission plan for as many of the National Marine Sanctuaries as possible includes three phases: (1) to provide the first photographic documentation of sanctuary plants, animals, and habitats at depths up to approximately 610 meters; (2) to expand characterization of habitats—focusing on larger animals such as whales, sharks, rays, and turtles—and compare habitat requirements among sanctuaries (Wilson, 1998); and (3) the all-important analysis, interpretation, and public dissemination of the masses of data collected.

The smallest, most remote, and least explored of the sanctuaries is the Fagatele Bay National Marine Sanctuary (FBNMS) in American Samoa, the only true tropical coral reef in the sanctuary system (figure 1). This site was largely unexplored below depths of approximately 30 meters until recently, and no comprehensive documentation of its plants, animals, and submarine topography exists. Little is known generally of shelf-edge coral reef habitats (50–100 meters deep) throughout the world, and no inventory of benthic-associated species has been compiled (e.g., Koenig et al., in press). FBMNS is also unique in that it is the only site with a submerged national park in the near vicinity. The National Park of American Samoa is also largely unexplored beyond the shallow coral reefs half a kilometer offshore. It will be extremely difficult to meet the sanctuary's and the park's mission of protecting the coral reef terrace and broader marine ecosystem without adequate knowledge of the deeper environment. Unlike the larger sanctuaries off the coast of the continental United States and Hawaii, the FBNMS will not be visited by the Deep-Worker submersible in the near future on an SSE mission, nor is the DeepWorker an adequate tool for surveying large regional areas. DeepWorker has been appropriate for the other sanctuaries because there already existed baseline surveys and maps on which to draw, so that the submersible could focus on specific regions to photograph and sample. Because FBNMS is so remote, there has been a critical need there for regional-scale, high-resolution, fully processed, interpreted, and accessible baseline data, in order to properly characterize the geological and biological environment.

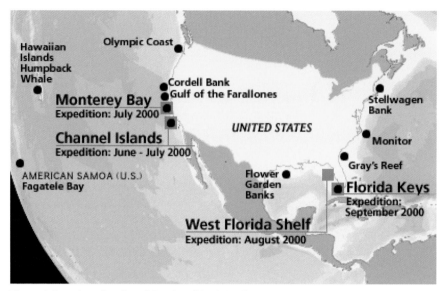

Figure 1. Location of twelve of the sites (black dots) comprising the National Marine Sanctuary System (excluding the most recent addition to the system at Thunder Bay in the Great Lakes region). Red squares indicate sites explored in the latter part of 2000 by the Sustainable Seas Expeditions. Map courtesy of the Sustainable Seas Expeditions and the National Geographic Society, www.nationalgeographic.com/seas.

GEOGRAPHICAL AND GEOLOGICAL SETTING

American Samoa (not to be confused with the independent nation of Samoa directly to the west) is the only U.S. territory south of the equator (figure 2) and is composed of five volcanic islands (from west to east: Tutuila, Anu'u, Ofu, Olosega, and Ta'u), as well as two small coral atolls, Rose and Swain (figure 3). Tectonically, the entire Samoan archipelago lies just east and 100 kilometers north of the subduction of the Pacific Plate beneath the northeastern corner of the Australian Plate at the Tonga Trench.

The estimated westward convergence rate of the Pacific Plate with respect to the Australian Plate along the entire length of the Tonga Trench is approximately 15 centimeters per year (Lonsdale 1986; DeMets et al. 1994). However, recent GPS measurements indicate an instantaneous convergence of 24 cm/yr across the northern Tonga Trench, which is the fastest plate velocity yet recorded on the planet (Bevis et al. 1995). This discrepancy in convergence rates appears to be related to seafloor spreading in the Lau Basin, west of the Tonga Trench. It has long been hypothesized that the islands of the Samoan archipelago were formed as a result of the tearing of the Pacific Plate as it turns abruptly to the west (aka "the Samoa corner") along the Tonga Trench (Isacks et al. 1969; Billington 1990). The Samoan chain is also unusual in that the islands are largest at the western end (Savai'i, Samoa), the middle is deeply eroded (Tutuila, American Samoa), and the easternmost feature (Rose Atoll, American Samoa) is a coral atoll that breaches the surface of the ocean, instead of an active underwater seamount (Hawkins and Natland 1975). In the Hawaiian archipelago, for instance, far to the north but oriented along a similar azimuth, these characteristics are completely reversed. The recent discovery, however, of the underwater volcano Vailulu'u to the east of the Samoan chain (Hart et al. 2000) provides strong evidence for a hotspot (as opposed to a "plate-tearing") origin for the islands, and one that is consistent with the westward plate movement of the Pacific.

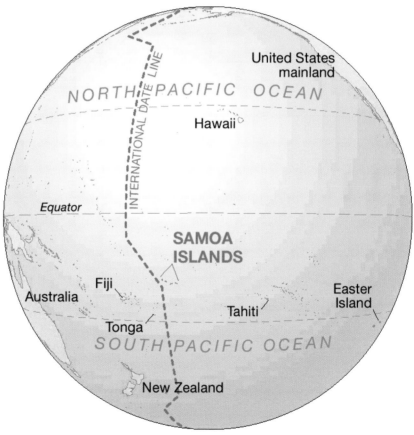

Figure 2. Regional map showing the location of the Samoan archipelago in the South Pacific.

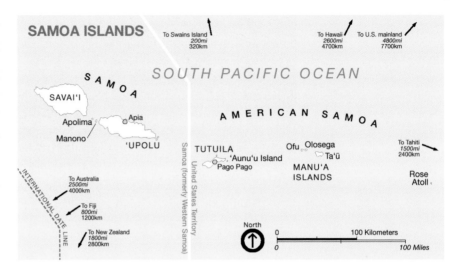

Figure 3. Regional map showing the islands of the independent nation of Samoa (formerly Western Samoa) to the west and American Samoa to the east (courtesy of the National Park of American Samoa, www.nps.gov/npsa/location.htm).

The FBNMS is located at the southwest corner of the island of Tutuila (figure 4). The bay is an ancient flooded volcano, with a thriving coral and calcareous algal reef community that is rapidly recovering from an infestation of crown-of-thorns starfish that devastated the corals in the late 1970s (Green et al. 1999). The bay was also pummeled by two hurricanes in 1990 and 1991, and a coral bleaching event occurred in 1994, possibly due to high sea-surface temperatures from an El Niño. Although much of the coral cover has been destroyed, fish populations still thrive, particularly surgeonfish, damselfish, and angelfish (Birkeland et al. 1987; Craig 1998). In addition, the steep slopes surrounding the bay contain some of the rarest paleotropical rain forests in the United States (www.fbnms.nos.noaa.gov). One of the greatest threats currently facing Fagatele Bay, as well as much of Samoa's coastal waters, is the rapid depletion of fish stocks by the illegal use of gill netting, spearfishing, poison, and dynamite (Sauafea, in press). In addition, the sanctuary staff is concerned about the potential for algal blooms with subsequent incidents of hypoxia (extremely low dissolved oxygen in the water) due to unchecked sewage outflow "upstream" from the bay.

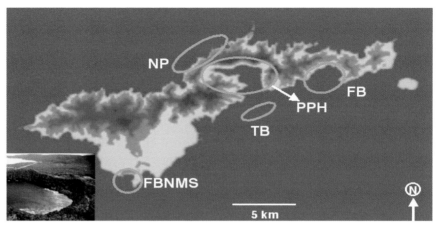

Figure 4. Index map of Tutuila, American Samoa, with pink circles showing the locations of recent multibeam bathymetric surveys around the island. Inset photograph at lower left is an aerial shot of the FBNMS (photo courtesy of the FBNMS, www.fbnms.nos.noaa.gov). Codes for other survey areas: NP = National Park of American Samoa (total area of submerged national park offshore Tutuila is ~5 square kilometers); PPH = Pago Pago Harbor; TB = Taema Bank; FB = Faga'itua Bay. Map is based on a U.S. Geological Survey (USGS) 10-meter digital elevation model (DEM) provided by A. Graves of Nuna Technologies, American Samoa.

SCIENTIFIC AND MANAGEMENT OBJECTIVES

Prior to the April–May 2001 mission, no scientific survey had been conducted in the deepest parts of Fagatele Bay (or in other marine regions around the territory). Two previous surveys reached depths of approximately 43 meters, but were both only brief, localized snapshots: an algal reconnaissance in 1996 (N. Daschbach 1996, unpublished data), and a rapid assessment survey for fish and coral in 1998 (Green et al. 1999). Therefore, two primary surveying objectives during the 2001 mission were to obtain (1) complete topographic coverage of the seafloor via a portable multibeam bathymetric mapping system, and (2) digital video and still photography of the biological habitats and physical features below 30 meters via advanced, mixed-gas diving equipment called rebreathers. Rebreathers overcome both the regulatory and physiological limitations of regular SCUBA, thereby making possible the establishment of coral reef transects as deep as 200 meters. In contrast to SCUBA, where the entire breath of a diver is expelled into the surrounding water when it is exhaled (open circuit), a rebreather apparatus is able to "reuse" the oxygen left unused in each exhaled breath (closed or semiclosed circuit), resulting in greatly extended dive times that are relatively quiet (little or no bubbles produced) and require much smaller tanks

(Elliot 2000). This chapter, however, focuses exclusively on the bathymetric mapping survey. Specific research questions that guided both surveys included:

- The primary question of what exactly is there. What is the character of the seafloor from a depth of 30 meters inshore out to the boundaries of the sanctuary (vertical relief, average depth, morphology and extent of reef structures, seafloor roughness, sand and/or algal cover, and such)? And further, what organisms and habitats currently reside within the sanctuary? Birkeland et al. (1987) and Green et al. (1997 and 1999) are the only existing reports on the first long-term, qualitative record of coral reef degradation in the region.

- What patterns are observed in the biological community structure within and beyond the reefs? What are the main physical parameters or mechanisms that are the causes of these structures (geometry or size of a bay; bathymetry; slope; percentage of coral, versus sand versus basalt; physical aftermath of illegal fishing or tropical storm impact; and so forth)?

- In 1996 visiting divers to the sanctuary reported unusually large fleshy algal blooms approximately 43 meters deep, suggesting a nutrient source in the bay that should be identified and monitored, particularly if it proves chronically harmful (Daschbach, unpublished data). Is it likely that the nutrient source is human-induced (e.g., sewage outfall carried to the bay along prevailing westerly currents or underground seepage into the bay from a landward watershed)? Where are the most appropriate sites for long-term monitoring of water quality and ocean currents?

- What are the broader implications for coral reef conservation and management (Gubbay 1995; Allison et al. 1998)? For example, which sites should be of special biological significance (such as a no-take zone or an area exhibiting a high degree of biological diversity)?

- Further questions may even be developed by local high school students or community college students as part of the ongoing educational outreach on the island conducted by the FBNMS staff (www.fbnms.nos.noaa.gov/HTML/Education.html).

**BASELINE SURVEY
DATA ACQUISITION**

Multibeam depth soundings for creating high-resolution seafloor maps were gathered by the Kongsberg Simrad EM-3000 system, contracted from the Center for Coastal Ocean Mapping of the University of South Florida (USF), and operated from a boat owned by the Department of Marine and Wildlife Resources (DMWR) of the American Samoa government (ASG; figure 5). The Kongsberg Simrad EM-3000 operates at a frequency of 300 kHz, fanning out up to 127 acoustic beams at a maximum ping rate of 25 Hz and at an angle of 130 degrees (1.5-by-1.5-degree beams are spaced 0.9 degrees apart). This yields swaths that are up to four times the water depth (figure 6). To ensure the accuracy of depth soundings across the full cross-track width of the swath, the EM-3000 combines them with critical information on both the attitude and position of the survey boat. Attitude is obtained with an inertial motion unit (IMU) from which adjustments can be made for the heave, roll, pitch, and heading of the boat, particularly through tight turns, rapid speed changes, and rough seas. The position of the boat is obtained via twenty-four-hour, precise-code (P-code) Global Positioning System (GPS) navigation. GPS fixes from two receivers aboard the boat were collected via with a POS/MV Model 320 (Position and Orientation System/Marine Vessels). At normal survey speeds of 3–12 knots the EM-3000 can capture depths in the 1- to 100-meter range (in warm, saline, equatorial waters), and data may be gridded at spacings of 0.25–25 meters. During our surveys, the accuracy of sounding was approximately 24 meters (i.e., points fell within a circle of radius 24 meters) and data was gridded at a resolution of 1 meter. With differential GPS (not available in real-time during our surveys), the system is capable of resolution down to the centimeter level, with a depth accuracy of 10–15 centimeters RMS, and a horizontal positional accuracy of less than 1 meter.

Figure 5. Thirty-foot survey boat of the American Samoa DMWR used for bathymetric surveys.

Figure 6. The transducer of the Kongsberg Simrad EM-3000 for transmitting and receiving acoustic beams. The transducer was mounted at the end of a metal pole secured to the port side pontoon of the survey boat and then cabled to the data collection and processing unit within the boat's cabin. The graphic at the right illustrates the swath emitted from standard multibeam transducers.

Another parameter essential to the accurate determination of depth is the velocity of sound, which changes according to the local temperature and salinity of the water (making correction of the resulting refraction of acoustic beams necessary). Sound velocity profiles were obtained for each survey around Tutuila with a sound velocimeter, deployed before the start of each sounding session.

Postprocessing steps began with tidal corrections made using verified tide data downloaded from NOAA. Preliminary pressure water level (N1) data used for the corrections was in meters above the mean lower low water (MLLW) datum obtained from reference station 1770000 in Pago Pago, American Samoa. Depth soundings were cleaned with a second-standard deviation filter, in order to flag outliers (i.e., points beyond two standard deviations of mean values within a moving window were flagged or removed).

Processed depth soundings from the EM-3000 system were available as ASCII xyz files. These were initially gridded using MB-Systems, a public-domain suite of software tools for processing and display of swath sonar data (Caress et al. 1996; www.ldeo. columbia.edu/ MB-System/html/mbsystem_home.html). Gridding was based on a Gaussian weighted average scheme, because it does the best job of representing a field of bathymetry in the absence of artifacts (Wright et al. 2000). The scheme is also heavily biased toward those data points closest to the center of a grid cell, thus minimizing anomalous values from outliers (Keeton et al. 1997). Each data point's contribution to a Gaussian weighted average for each nearby grid cell was calculated as the point was read and added to the grid cell sums (after the method of Caress and Chayes, 1995). Gaps between swaths were filled using a thin plate spline (i.e., a common smoothing function; see Sandwell 1987) with a tension of infinity. The clipping dimension for the spline interpolation was increased by varying distances, up to five times the grid spacing, in order to fill data gaps.

MB-Systems outputs grids in the format of Generic Mapping Tools (GMT; gmt.soest.hawaii.edu) a public-domain, open-source collection of UNIX® tools for manipulating data sets and producing maps and illustrations, and a de facto standard within the academic multibeam swath mapping and marine geology/geophysics communities. Grids were then converted to ArcInfo format with ArcGMT, a public-domain suite of tools for converting GMT-style grids to Arc format (Wright et al. 1998; dusk.geo.orst.edu/arcgmt). Final spacings for most grids were 1 meter, with coordinates in latitude/ longitude decimal degrees in the WGS84 datum.

RESULTS AND DISCUSSION

BATHYMETRIC SURVEYS

Maps created with data from the Kongsberg Simrad multibeam system are particularly good at showing the details of morphological features. Many of the cookie-cutter bays that are found along the southern coast of Tutuila, such as Fagatele Bay, Larsen Bay, Pago Pago Harbor, and Faga'itua are thought to be the result of volcanic collapse and erosion (Stearns 1944). As crustal loading caused the island to subside, large portions of these eroded valleys were flooded by the sea. Figure 7 shows complete bathymetric coverage of the FBNMS embayment, in which a primary reef platform is delineated around the inner (shoreward) rim of the bay out to approximately 15–20 meters of depth. Of note are two slender peninsulas, likely erosional remnants, extending from the main platform at 14° 21′ 47″ S, 170° 46′ W and 14° 22′ S, 170° 45′ 53″ (figure 7). The second peninsula was noted and explored by the rebreather divers, and found to be an important habitat where several new species of fish were discovered (described more fully below). Below the main platform are well-developed terraces that slope down to approximately 80 meters; they are likely the record of platforms cut by sea-level oscillations late in the history of the volcanic edifice. A reef terrace at 14° 22′ S cuts the bay into two provinces: a shoreward, circular depression that extends to a maximum depth of 145 meters, and a more elongated depression seaward, extending down to a maximum depth of 160 meters. The reef terrace at 14° 22′ S rises 50 meters from the depths of the bay; it may have formed at sea level on the outer margins of the volcanic edifice and was subsequently drowned as the island subsided below sea level.

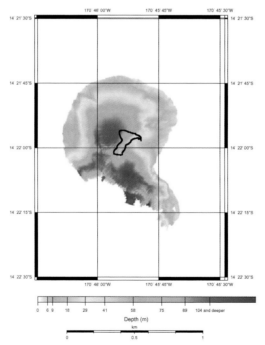

Figure 7. Color-shaded bathymetric map of the FBNMS, created from a 1-meter Kongsberg Simrad EM-3000 grid (see figure 4 for regional location). Color change interval based on a histogram equalization of the bathymetry to accentuate morphological features. Solid line delineates the estimated dive track of a rebreather diving mission in the sanctuary, immediately following bathymetric surveying (Pyle 2001). Map projection is Mercator.

What will be important for future surveys of this and other embayments around Tutuila is the ability to characterize the nature of outcrops, terraces, platforms, and floors with backscatter imagery, in addition to the bathymetry. (The software necessary for EM-3000 backscatter processing and classification was not available to the authors for these initial surveys.) For instance, Exon (1982) reports that the shallow flanks of the Samoan islands are characterized by outcrops of basalt and limestone, biogenic and volcanic silt, sand and gravel, calcareous pavements and calcareous ooze. Building upon this earlier study, recharacterization of the seafloor with the newer, higher-resolution sidescan sonar units will be particularly useful for mapping benthic habitats and species distributions. Acoustic backscatter imagery provides a better means than bathymetry alone for evaluating habitat characteristics such as sediment and algal cover, unchannelized debris, slump sheets, and so on, and determining what attracts and retains echinoderms, fishes, sea turtles, and perhaps even marine mammals. The imagery produces fairly sharp boundaries, with pixels easily classed into different categories. For instance, shelf-edges on the seafloor may be high in relief (up to 20 meters above the seafloor) or low in relief (less than a meter above the seafloor), each type indicative of a possible habitat. However, although geomorphological classifications may be straightforward, assemblages of benthic organisms and associated substrata are more difficult to classify because they often exhibit considerable variation and tend to grade gradually from one assemblage to another, requiring direct visualization for identification and estimation of cover (Richards 1997; Coleman et al. 1999). And finally, it will be interesting to note whether submarine features are of a constructional origin (e.g., volcanic or volcaniclastic cones) or whether they are the result of slumps or debris slides. Keating et al. (2000) have found that mass wasting plays an important role in the geologic development of volcanic edifices on the nearby Samoan island of Savai'i, and that it may be an important component of the geologic record for all of the Samoan islands.

As mentioned earlier, the new bathymetry of the FBNMS also helped to guide the location of a deep-diving mission to the sanctuary on May 16, 2001, immediately following the bathymetric surveys around Tutuila (figure 7). Divers used rebreathers to work underwater for more than three and a half hours (a block of time significantly longer than allowed by traditional SCUBA), collecting videotape of coral reef biota and habitats up to a maximum depth of 113 meters (Pyle 2001).

Even though the diving mission was cut short by poor weather, divers observed twelve completely new species of fish, seventeen species that had never been observed in American Samoa, and several species that were previously unknown to the waters of Fagatele Bay (Pyle 2001). For instance, figure 8a shows a reddish fish that is a new (as yet unnamed) species of the genus *Cirrhilabrus,* discovered recently during a similar deep-diving expedition to Fiji. Figure 8b shows what is thought to be a new species of sand perch, in the genus *Parapercis.* The species is similar to another new species in the same genus from deep reefs in Papua, New Guinea, but has never before been observed in the South Pacific. Processing of video transect data from the dive is ongoing, and upon completion the dive track, observation attributes, and hotlinked images will be added to the FBNMS GIS.

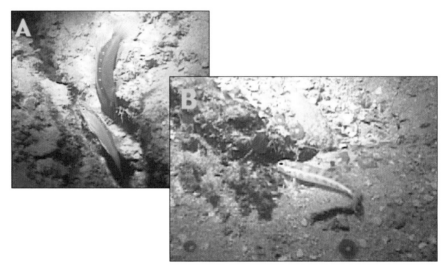

Figure 8. Still images captured from Sony VX-1000 videotape footage taken during a rebreather dive at the FBNMS on May 16, 2001 (Pyle 2001). (a) Two individuals of a new fish species of genus *Cirrhilabrus,* family *Labridae,* only recently discovered on deep reefs in Fiji and never before seen in American Samoa. Image taken at a depth of 113 meters, dive time 00:28. (b) A newly discovered species of sand perch, of the genus *Parapercis,* family *Pinguipedidae,* never before seen in the South Pacific. Image taken at a depth of 82 meters, dive time 00:33.

The FBNMS has had long-standing partnerships with many other government agencies on the island, such as the aforementioned DMWR, the ASG's Department of Commerce (DOC), the National Marine Fisheries Service (NMFS), the American Samoa Coastal Management Program (ASCMP), the Historic Preservation Office, and the National Park of American Samoa. Figures 9 through 12 show bathymetric data that will be of interest not only to the FBNMS, but to these agencies as well, with the data to be shared and integrated via the FBNMS GIS. Figure 9 shows bathymetric coverage of a portion of the National Park of American Samoa, from Fagasa Bay to Tafeu Cove on the north coast of Tutuila. The map shows the first complete bathymetric coverage of a drowned barrier reef, out to a depth of approximately 15 meters, a fairly broad reef shelf at around 30 meters, and the shelf-edge break near 40–50 meters. Recent studies by National Park Service scientists (Craig 1996 and 1998; Craig et al. 1997) summarize the population biology, spawning patterns, and subsistence harvesting of the most abundant surgeonfish species that frequent the shallow portion of the reef, the alogo (*Acanthurus lineatus*), the manini (*Acanthurus triostegus*), and the pone (*Ctenochaetus striatus*). Other submerged areas of the national park yet to be surveyed include barrier reefs off the southeast shore of Ofu, and off the south and east shores of Ta'u, both islands nearly 100 kilometers to the east of Tutuila. Very little is known of these environments beyond the shelf-edge break at 50 meters.

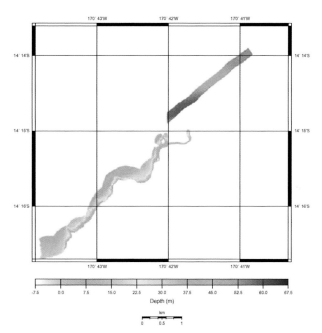

Figure 9. Color-shaded bathymetric map of a portion of the National Park of American Samoa along the north central coast of Tutuila, created from a 2-meter Kongsberg Simrad EM-3000 grid (see figure 4 for regional location). Map projection is Mercator.

Figure 10a shows complete bathymetric coverage of Pago Pago Harbor, another site of large caldera collapse, erosion, and subsidence. Out to a depth of 10–15 meters, the harbor is lined with fringing reefs and dotted with small knolls that could be blocks of reef debris or patch reefs. Green et al. (1997) note that reefs here account for one-fifth of all coral reefs on the island, but unfortunately have been severely degraded over the past eight decades due mainly to two hurricanes and a major coral-bleaching event in the 1990s, but also due to pollution from tuna canneries, dredging and filling of the reef flat, and ship traffic. Indeed, the high resolution of the EM-3000 enabled the delineation of a major shipwreck in the harbor (figure 10b). The USS *Chehalis* was a World War II oil and gas tanker that exploded and sank in the harbor in 1949. The wreck is 90 meters long, lies in 45 meters of water, and is thought to still be a source of pollution affecting water quality.

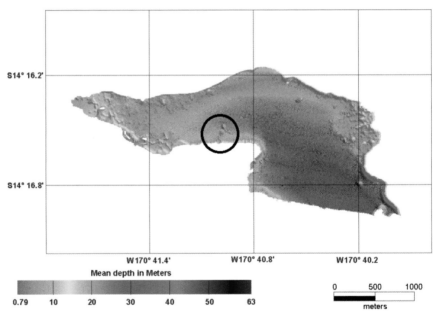

Figure 10a. Color-shaded, sun-illuminated bathymetric map of Pago Pago Harbor, the largest deepwater harbor in the South Pacific (see figure 4 for regional location). Map created from a 1-meter Kongsberg Simrad EM-3000 grid in the Mercator projection. Circle shows the location of a major shipwreck, detailed in (b).

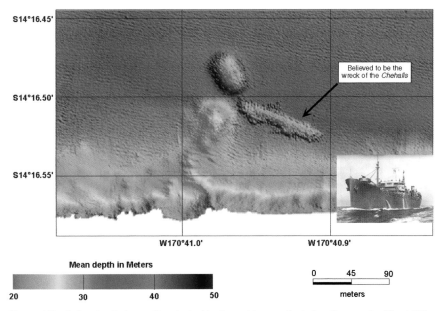

S14°16.45'

S14°16.50'

S14°16.55'

Believed to be the
wreck of the *Chehalis*

W170°41.0' W170°40.9'

Mean depth in Meters

20   30   40   50

0   45   90

meters

Figure 10b. Color-shaded, sun-illuminated bathymetric map featuring the wreck of the USS *Chehalis* in Pago Pago Harbor. Map created from the same 1-meter grid as in (a), in the Mercator projection. Inset photo of a ship in the same class as the USS *Chehalis* courtesy of the U.S. National Archives and Records Administration, www.nara.gov/publications/sl/ navyships/auxil.html.

Moving northeast along the coast from Pago Pago Harbor is Faga'itua Bay, the bathymetry of which is shown in figure 11. A reef shelf out to approximately 20 meters deep is very broad and prominent, much more so than any of the other sites surveyed. And at the base of the shelf are four prominent outcrops, two that are roughly circular and 75–100 meters in diameter, and two that are roughly linear and 300 meters long. These features may be blocks of reef debris or patch reefs, and the circular ones may be volcanic or volcaniclastic cones such as those observed by Keating et al. (2000) onshore along the northeast coast of Savai'i and in the Manua island group (Ofu, Olesega, and Ta'u).

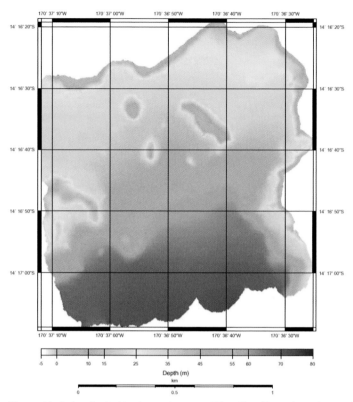

Figure 11. Color-shaded bathymetric map of Faga'itua Bay, along the southeast coast of Tutuila, created from a 1-meter Kongsberg Simrad EM-3000 grid (see figure 4 for regional location). Map projection is Mercator.

Taema Bank (figure 12) is a long, narrow submarine platform approximately 3 kilometers long by 30 meters wide that rises some 30 meters above the surrounding seafloor. The abyssal seafloor in the vicinity of the bank is nearly 110 meters deep. Because the platform is flat and smooth, it is thought to be an ancient reef terrace that may have once experienced wave erosion at sea level. According to Stearns (1944) and Flanigan (1983), Taema Bank has a geological connection to the caldera that collapsed and subsided to form what is now Pago Pago Harbor. They note how the southerly tilt of the caldera, the slope of the caldera fill, and the sizes and shapes of the volcanics governed the course of the prehistoric Pago Pago River. The river once eroded a deep valley along the northern and eastern caldera rim, and now, due to the subsidence of the island, can be traced out to sea as far as Taema Bank, where it has been obscured by subsequent coral growth (Stearns 1944; Flanigan 1983).

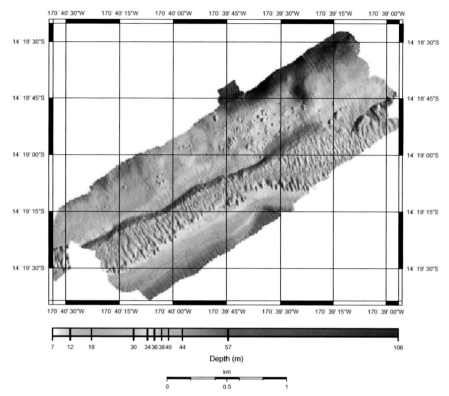

Figure 12. Histogram equalized, shaded relief bathymetric map of Taema Bank, located ~3 kilometers off the south central coast of Tutuila (see figure 4), created from a 1-meter Kongsberg Simrad EM-3000 grid. The bathymetry is illuminated at an azimuth of 270 degrees using a shading magnitude of 0.4 to accentuate the northwest trending bank. Color change interval based on histogram equalization. Map projection is Mercator.

GIS COORDINATION

Coincident with the bathymetric surveying was the conversion and integration of bathymetric grids into a new GIS for the FBNMS (figure 13). Included also was a compilation of mostly terrestrial data layers (DEMs, DLGs, digital raster graphic (DRG) files, shapefiles, coverages, and grids) obtained from the National Park Service, the USGS, the Digital Chart of the World, and other sources. As most of the data was undocumented, FGDC-compliant metadata records were prepared for each layer in the compilation, as well as for the bathymetric grids, using the NOAA Coastal Services Center metadata collector tool, version 2 (CD–ROM insert, this volume; figure 14).

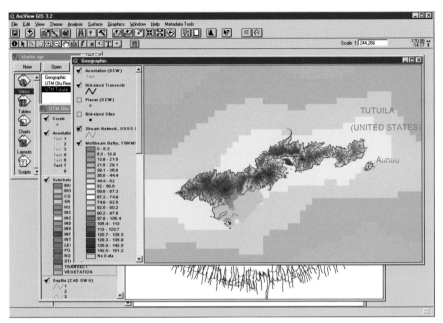

Figure 13. Screen capture of the new FBNMS GIS including all multibeam bathymetry grids, as well as compilation of terrestrial GIS data layers obtained from other sources. The top view shows Smith and Sandwell (1997) regional bathymetry (in shades of aqua), with a USGS 10-meter DEM for the main island of Tutuila, overlain with a stream network based on USGS digital line graph (DLG) data, as well as bathymetric grids for FBNMS and Pago Pago Harbor.

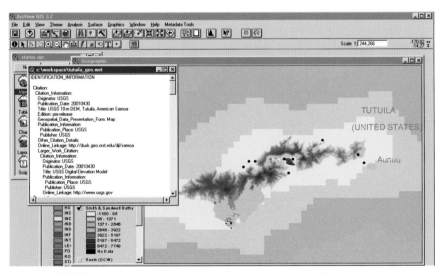

Figure 14. FGDC-compliant metadata records were written for all grids, coverages, and shapefiles in the FBNMS GIS using the NOAA Coastal Services Center metadata collector tool (see CD–ROM insert).

Because of the lack of differential GPS positioning for the bathymetry, some grids were initially compared to the USGS 10-meter DEM of the island of Tutuila (1:24,000 in UTM Zone 2, WGS84), in order to roughly judge positional accuracy (i.e., did the shoreward edges of the bathymetry grids line up reasonably with the shoreline of the USGS DEM?). Experimenting with the Tutuila DEM as a reference layer, copies of grids were successfully shifted using the SHIFT function in ArcGrid™. In addition, a stream network coverage—based on a USGS DLG in an antiquated local datum known as the American Samoa Datum of 1962 (ASD62)—required a substantial amount of coordinate transformation (via several sessions in ArcCONTROLPOINTS to create coordinate links between the reference DEM in UTM and a copy of the original coverage that had been projected from ASD62 to UTM, WGS84), and then rubbersheeting in ArcADJUST. After a final step of removing dangling arcs and nodes in ArcEdit™, the stream network coverage matched quite well with the Tutuila DEM.

Indeed, many of American Samoa's terrestrial data sets are stored as State Plane coordinates under ASD62, a standard that has been in use by the territorial government for several years. It has been extremely difficult for local users to migrate data to higher levels of accuracy (e.g., workable transformations to UTM Zone 2, WGS84). Fortunately, the datum issue should soon be resolved. At the time of this writing, the National Geodetic Survey announced plans to establish a new continuously operating reference station (CORS) on Tutuila that will provide GPS carrier phase and code range measurements for updating the survey grid and supporting future 3-D positioning activities on the island (A. Graves, pers. comm.). This and many other projects are under the purview of a fairly new American Samoa GIS User Group, comprised of thirty representatives of local and federal agencies, along with the American Samoa Community College, in an effort to develop GIS standards for the territory. In addition to establishing map projection and datum standards, other major challenges for the group include finding effective ways of exchanging and integrating data among and outside of agencies, acquiring high-resolution satellite imagery for planning and mapping purposes, and building a local infrastructure that will maintain and update GIS hardware, software, databases, and project initiatives.

In order to make all data sets in the FBNMS GIS accessible, not only to the sanctuary staff and their collaborators in American Samoa but to collaborators throughout Oceania and the United States, a Web clearinghouse was built at dusk.geo.orst.edu/djl/samoa (figure 15). The site provides links to all of the GIS data and metadata, and to bathymetric grids in GMT format, various maps, photographic images, and graphics. All GIS data is provided as ArcInfo export interchange files (i.e., *.e00 files), which may be imported into ArcInfo, ArcView, or ArcExplorer™.

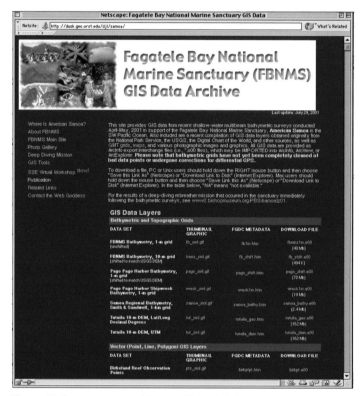

Figure 15. Screen capture of the new FBNMS GIS Web site at dusk.geo.orst.edu/djl/samoa, with free downloads of raster and vector GIS data, images, maps, thumbnail graphics, and FGDC-compliant metadata.

The vision for long-term coordination of the FBNMS GIS includes: (1) development of protocols and maintenance procedures for future acquisitions of data; (2) establishing an order of prioritization for GIS data integration based both on availability and quality of the data; (3) timely integration of that data into the GIS, using simple format filters and data input programs for ArcView and ArcInfo as necessary (e.g., Wright et al. 1998 and CD–ROM insert, this volume; Wong et al. 1999); (4) continued documentation of data sources and contact information, preferably with the NOAA CSC metadata collector tool, to create FGDC-compliant metadata; and (5) continued uploading to the Web-based clearinghouse where users can view and query data and metadata online.

A fortuitous occurrence at the time of bathymetric surveying and GIS activities for the FBNMS was the visit of a NOAA Coastal Services Center delegation, the purpose of which was to involve American Samoa in a new Pacific Islands GIS initiative. The project, begun in April 2001, is a multiyear initiative to build sustainable spatial data capacity within the coastal resource management programs of Hawaii and the U.S. territories of American Samoa, Guam, and the Commonwealth of the Northern Mariana Islands, as well as to leverage other related federal activities in the region (C. Fowler, pers. comm.). Members of the NOAA delegation discussed ways in which the initiative could aid the American Samoa User Group in obtaining additional GIS training, equipment, software, data, assistance with geodetic control issues, and the placement of interns to assist with GIS data input and analysis.

CONCLUSION

GIS capacity building for the FBNMS has been extremely successful in terms of baseline data acquisition, the setup of a Windows NT® workstation dedicated to GIS, and the construction of an ArcView project document with paths to more than twenty raster and vector themes and their associated metadata, tables, images, and Avenue™ scripts. Surveys with a portable multibeam bathymetric mapping system yielded complete coverage of previously unexplored regions of the FBNMS, as well as many other sites around the island of Tutuila. The data was of excellent quality and guided a ground-truthing dive mission into the FBNMS to observe and videotape fauna inhabiting regions below 45 meters. GIS grids created from the bathymetric surveys will provide base layers for the purposes of visual overlay and comparison with other data sets in American Samoa. And the new Web clearinghouse of all data, bathymetric and terrestrial, will facilitate the continued distribution and synthesis of information in support of the FBNMS, with the additional hope of increasing opportunities for research and managerial collaboration.

Initial data integration has been an important first step for the sanctuary, but spatial analysis will certainly play an important part in answering key research and management questions, and in establishing survey, sampling, monitoring, and management protocols. For example, Bridgewater (1993) and Aspinall (1995) note that combining a landscape ecology approach (i.e., data analysis guided by purposeful ecological objectives) with a GIS is desirable because it allows for the study of structure, function, and change within ecosystems (such as coral reefs), while attempting to manage the many spatial and temporal scales. For the April–May 2001 survey, a primary long-term objective is to analyze physical factors important to coral reef development in FBNMS, such as habitat classification, submarine aspect, submarine slope, and bottom substrate relief, along with several community descriptors, via GIS query, spatial correlation tests, and buffer analysis. Treml (1999) was successful with this approach in analyzing coral reef community ecology on St. John, U.S. Virgin Islands, using factors such as current regime, substrate characteristics, coastal topography, bay geometry, watershed size, sedimentation, tropical storm impact, bathymetry, biodiversity, evenness of biota distribution, and algae cover.

A vision for subsequent GIS spatial analyses therefore includes:

- Creation of coastal terrain models (after Li et al., in press) consisting of terrestrial DEMs merged with multibeam bathymetric grids, likely at a 10-meter resolution, with interpolation of any gaps attained with an inverse distance weighted (IDW) algorithm, and derivation of slope and aspect from these grids.

- As bathymetric relief will likely be a potentially important factor for biological community distribution, the next step will be to generate indices of relief, after the following method of Bushing (1995). The aspect layer may be reclassed into several degree categories (e.g., 0–14 degrees of view). The resulting image may then be imported into image-processing software, and rectangular diversity or texture (Star and Estes 1990) filters will be applied at several spatial resolutions to model the heterogeneity in the region of each raster cell. The variable dimensions of the spatial filters will permit the investigation of bottom relief at three separate spatial scales. For instance, 3×3, 5×5, and 10×10 pixel filters should yield an index of bottom relief at spatial scales of about 30–40, 50–70, and 100–140 meters respectively from the subject cell.

These techniques were first used to quantify terrestrial landscape characteristics for determining measures of landscape heterogeneity, habitat diversity, and habitat fragmentation (Weaver and Kellman 1981; Ripple et al. 1991), and hold similar potential for undersea environments.

• Use of graphical and analytical overlay (i.e., union, intersect, join functions) to relate biological distribution patterns to seafloor parameters such as depth, submarine aspect, submarine slope, and bottom substrate indices of relief, as well as sediment cover and lava flow morphology. Use of correlations to characterize regions of biological diversity, habitat classes, areas requiring special protection such as no-take, etc.

A final outcome of this project was the presentation of the results of the bathymetric surveying and GIS efforts, as well as a discussion of the utility of GIS for coral reef studies and sanctuary management, as part of an SSE virtual teacher workshop entitled "Conservation and the Coral Reef World." There was also a separate breakout room, hosted by sanctuary manager Nancy Daschbach, devoted to discussion of some of the complex processes that occur in tropical coral reefs, and development of an understanding of the challenges facing coral reef resource management in remote areas. The workshop was held in the summer and fall of 2001, completely over the Web (www.coexploration.org/sse), for several hundred enrollees worldwide, using a conferencing software package called Caucus. SSE virtual teacher workshops take advantage of Web-based and interactive television technology to reach large numbers of K–12 teachers, particularly those from noncoastal regions or from traditionally underrepresented minority groups (Cava 2001). The SSE has just begun to employ these as innovative ways to provide teachers with direct access to SSE scientists and collaborators, and to encourage them to work ocean studies into their teaching practices. Teachers also receive instruction on important Internet skills such as effective use of online search tools, working with online images, and accessing and analyzing online data and maps (Cava 2001).

ACKNOWLEDGMENTS

The authors would like to acknowledge the excellent support and contributions of Nancy Daschbach, manager of the Fagatele Bay National Marine Sanctuary. We are particularly grateful to Captain Terry Lam Yuen and his intrepid crew (Tusi, Lautofa Lutu, and Palala Pule) for excellent service both in port and at sea, as well as to the American Samoa DMWR for the use of their boat. Thanks also to John McDonough of the NOAA National Ocean Service Office of Special Programs, Richard Pyle of the Bernice Pauahi Bishop Museum, and Francesca Cava of SSE. Special thanks to Allison Graves of Nuna Technologies for providing USGS DEM, DLG, and DRG data sets, as well as several fruitful discussions. BTD and DFN were contracted by the FBNMS via the U.S. Department of Commerce, Western Administration Support Center, NOAA Pacific Marine Environmental Laboratory, Seattle, Washington, and acknowledge multibeam support by the Office of Naval Research and the USGS. DJW was supported by National Science Foundation grant OCE/EHR-0074635.

REFERENCES

Aspinall, R. J. 1995. Geographical information systems: Their use for environmental management and nature conservation. *Parks* 5(1):20–31.

Allison, G. W., J. Lubchenco, and M. H. Carr. 1998. Marine reserves are necessary but not sufficient for marine conservation. *Ecological Applications* 8(1):S79–S92.

Bevis, M., F. W. Taylor, B. E. Schultz, J. Recy, B. L. Isaacks, S. Helu, R. Singh, E. Kendrick, J. Stowell, B. Taylor, and S. Calmant. 1995. Geodetic observations of very rapid convergence and back-arc extension at the Tonga arc. *Nature* 374:249–51.

Billington, S. 1990. *The Morphology and Tectonics of the Subducted Lithosphere in the Tonga-Fiji-Kermadec Region from Seismicity and Focal Mechanism Solutions.* Ph.D. thesis. Ithaca, New York: Cornell University.

Birkeland, C. E., R. H. Randall, R. C. Wass, B. Smith, and S. Wilkins. 1987. *Biological Assessment of the Fagatele Bay National Marine Sanctuary.* NOAA Technical Memorandum.

Bridgewater, P. B. 1993. Landscape ecology, geographic information systems and nature conservation. In *Landscape Ecology and Geographic Information Systems,* eds. R. Haines–Young, D. R. Green and S. Cousins, 23–36. London: Taylor & Francis.

Bryant, D., L. Burke, J. McManus, and M. Spalding. 1998. *Reefs at Risk: A Map-Based Indicator of Potential Threats to the World's Coral Reefs.* Washington, D.C.: World Resources Institute.

Bunce, L. L., J. B. Cogan, K. S. Davis, and L. M. Taylor. 1994. The National Marine Sanctuary Program: Recommendations for the program's future. *Coastal Management,* 22:421–26.

Bushing, W. W. 1995. Identifying regions of persistent giant kelp (*Macrocystis pyrifera*) around Santa Catalina Island for designation as marine reserves. *Proceedings of the 15th Annual ESRI User Conference,* Paper 247. Palm Springs, California.

Caress, D. W., and D. N Chayes. 1995. New software for processing sidescan data from sidescan-capable multibeam sonars. *Proceedings of the IEEE Oceans 95 Conference,* 997–1000.

Caress, D. W., S. E. Spitzak, and D. N. Chayes. 1996. Software for multibeam sonars. *Sea Technology* 37:54–57.

Cava, F. M. 2001. The Sustainable Seas Expeditions: A National Geographic Society, National Oceanic and Atmospheric Administration and Richard and Rhoda Goldman Fund project for ocean exploration and public education. *Proceedings of The Future is Here: A Conference for Environmental Education.* Royal Melbourne Institute of Technology, Melbourne, Australia.

Coleman, F. C., C. C. Koenig, A. M. Eklund, and C. B. Grimes. 1999. Management and conservation of temperate reef fishes in the grouper–snapper complex of the southeastern United States. In *Life in the Slow Lane,* ed. J. A. Musick. American Fisheries Society Symposium 23:233–42.

Craig, P. 1996. Territoriality and time-budget of the surgeonfish *Acanthurus lineatus* in American Samoa. *Environmental Biology of Fishes* 46:27–36.

Craig, P. 1998. Temporal spawning patterns for several surgeonfishes and grasses in American Samoa. *Pacific Science* 52:35–39.

Craig, P., et al. 1997. Population biology and harvest of a coral reef surgeonfish (*Acanthurus lineatus*) in American Samoa. *Fishery Bulletin* 95:680–93.

DeMets, C., R. G. Gordon, D. F. Argus, and S. Stein. 1994. Effect of recent revisions to the geomagnetic reversal time scale on estimates of current plate motions. *Geophysical Research Letters* 21 (20):2,191–94.

Earle, S. A., and W. Henry. 1999. *Wild Ocean: America's Parks Under the Sea.* Washington, D.C.: National Geographic Society.

Elliott, B. 2000. *The Rebreather Web Site.* Northwood Designs, Inc. Antwerp, New York: www.nwdesigns.com/rebreathers.

Exon, N. F. 1982. Offshore sediments, phosphorite and manganese nodules in the Samoan region, Southwest Pacific. *South Pacific Marine Geological Notes,* SOPAC Technical Report 103–20.

Flanigan, J. M. 1983. *Earth, Fire, and Water,* unpublished report, American Samoa Community College: naio.kcc.hawaii.edu/jflanigan/EFW/theisland.html.

Green, A. L., C. E. Birkeland, R. H. Randall, B. D. Smith, and S. Wilkins. 1997. 78 years of coral reef degradation in Pago Pago Harbor: a quantitative record. *Proceedings of the 8th International Coral Reef Symposium.* Panama City, Panama, 2:1,883–88.

Green, A. L., C. E. Birkeland, and R. H. Randall. 1999. Twenty years of disturbance and change in Fagatele Bay National Marine Sanctuary, American Samoa. *Pacific Science* 53(4):376–400.

Gubbay, S., ed. 1995. *Marine Protected Areas: Principles and Techniques for Management,* London: Chapman and Hall.

Hart, S. R., H. Staudigel, A. A. P. Koppers, J. Blusztajn, E. T. Baker, R. Workman, M. Jackson, E. Hauri, M. Kurz, K. Sims, D. J. Fornari, A. Saal, and S. Lyons. 2000. Vailulu'u undersea volcano: The new Samoa. *Geochemistry, Geophysics, Geosystems* 1: Paper number 2000GC000108. Electronic journal: g-cubed.org

Hawkins, J. W., Jr. and J. H. Natland. 1975. Nephelinites and basanites of the Samoan linear volcanic chain: Their possible tectonic significance. *Earth and Planetary Science Letters* 24:427–39.

Isacks, B. L., L. R. Sykes, and J. Oliver. 1969. Focal mechanisms of deep and shallow earthquakes in the Tonga–Kermadec region and the tectonics of island arcs. *Geological Society of America Bulletin* 80:1,443–70.

Jones, G. P., M. J. Milicich, et al. 1999. Self-recruitment in a coral reef fish population. *Nature* 402:802–804.

Keating, B. H., C. E. Helsley, and I. Karogodina. 2000. Sonar studies of submarine mass wasting and volcanic structures off Savai'i Island, Samoa. *Pure and Applied Geophysics* 157(6–8):1,285–313.

Keeton, J. A., R. C. Searle, B. Parsons, R. S. White, B. J. Murton, L. M. Parson, C. Peirce, and M. C. Sinha. 1997. Bathymetry of the Reykjanes Ridge. *Marine Geophysical Researches* 19, 55–64.

Koenig, C., F. Coleman, G. Fitzhugh, C. Gledhill, et al. In press. Marine reserves for the protection of critical shelf-edge spawning habitat for economically important reef-fish. *Bulletin of Marine Science.*

Li, R., R. Ma, and K. Di. In press. Digital tide-coordinated shoreline. *Marine Geodesy.*

Lonsdale, P. 1986. A multibeam reconnaissance of the Tonga Trench axis and its intersection with the Louisville Guyot Chain. *Marine Geophysical Researches* 8:295–327.

Pyle, R. 2001. *B. P. Bishop Museum Exploration and Discovery: The Coral-Reef Twilight Zone, Fagatele Bay National Marine Sanctuary,14–18 May 2001*, Bernice Pauahi Bishop Museum, Honolulu, Hawaii, www2.bishopmuseum.org/PBS/samoatz01.

Richards, W. J. 1997. Site characterization for the Florida Keys National Marine Sanctuary and environs. *Bulletin of Marine Science* 61(2):505–10.

Ripple, W. J., G. A. Bradshaw, and T. A. Spies. 1991. Measuring forest landscape patterns in the Cascade Range of Oregon, USA. *Biological Conservation* 57:73–88.

Sandwell, D. T. 1987. Biharmonic spline interpolation of GEOS-3 and Seasat Altimeter data. *Geophysical Research Letters* 14:139–42.

Sauafea, F. S. In press. Community-based fisheries management in American Samoa. *Proceedings of the Fifth Regional Symposium, PACON 2001.* Burlingame, California: Pacific Congress on Marine Science and Technology.

Smith, W. H. F. and D. T. Sandwell. 1997. Global seafloor topography from satellite altimetry. *Science* 277:1,957–62.

Star, J. and J. Estes. 1990. *Geographic Information Systems: An Introduction.* Englewood Cliffs, New Jersey: Prentice Hall.

Stearns, H. T. 1944. Geology of the Samoan islands. *Geological Society of America Bulletin* 55:1,270–332.

Treml, E. 1999. *Fringing Reef Framework Development and Maintenance of Coral Assemblages along St. John's South Shore: A Geographic Information System (GIS) Analysis.* Master's thesis, University of Charleston, Charleston, South Carolina.

Weaver, M. and M. Kellman. 1981. The effects of forest fragmentation on woodlot tree biotas in Southern Ontario. *Journal of Biogeography* 8:199–210.

Wilson, A. D. 1998. America's fragile sea sanctuaries: Exploring and documenting. *Sea Technology* 39:29–33.

Wolanski, E., ed. 2001. *Oceanographic Processes of Coral Reefs: Physical and Biological Links in the Great Barrier Reef.* Boca Raton, Florida: CRC Press.

Wong, F. L., S. L. Eittreim, C. H. Degnan, and W. C. Lee 1999. USGS Seafloor GIS for Monterey Sanctuary: Selected data types. *Proceedings of the 19th Annual ESRI User Conference*, Paper 659. San Diego, California.

Wright, D. J., R. Wood and B. Sylvander. 1998. ArcGMT: A suite of tools for conversion between ARC/INFO® and Generic Mapping Tools (GMT). *Computers and Geosciences* 24(8):737–44.

Wright, D. J., S. H. Bloomer, C. J. MacLeod, B. Taylor, and A. M. Goodliffe. 2000. Bathymetry of the Tonga Trench and forearc: A map series. *Marine Geophysical Researches* 21(5):489–511.

ABOUT THE AUTHORS

Dawn J. Wright, Ph.D., is an associate professor in the Department of Geosciences at Oregon State University, where she has been on the faculty since 1995. Prior to joining the faculty there, she was a seagoing marine technician for the international Ocean Drilling Program, a graduate research assistant for the Marine Science Institute and the National Center for Geographic Information and Analysis (NCGIA) at the University of California, Santa Barbara (UCSB), and a postdoctoral research associate at the NOAA Pacific Marine Environmental Laboratory in Newport, Oregon. A few years after the deepsea vehicle Argo I was used to discover the *Titanic* in 1986, Dawn was presented with some of the first GIS data sets to be collected with that vehicle. It was then that she first became acutely aware of the challenges of applying GIS to deep marine environments. She has since completed oceanographic fieldwork (often with GIS) in some of the most geologically active regions on the planet, including the East Pacific Rise, the Mid-Atlantic Ridge, the Juan de Fuca Ridge, the Tonga Trench, and volcanoes under the Sea of Japan and the Indian Ocean. Her research interests include application and analytical issues in GIS for oceanographic data (many of which are being tackled with graduate students in her lab, Davey Jones' Locker, dusk.geo.orst.edu/djl); the relationships between volcanic, hydrothermal, and tectonic processes at seafloor-spreading centers; the analysis and interpretation of data from deepsea mapping systems; and the geography of cyberspace. Dawn is also coeditor of the book *Marine and Coastal Geographical Information Systems* (Taylor & Francis, publishers, 1999/2000). She holds a bachelor of science (cum laude) in geology from Wheaton College (Illinois), a master of science in oceanography from Texas A&M, and a doctoral degree in physical geography and marine geology from UCSB.

Brian T. Donahue is a research assistant at the Center for Coastal and Ocean Mapping at the University of South Florida (USF) where he coordinates several geophysical research activities, including the setup, repair, acquisition, and processing of the Kongsberg Simrad EM-3000 multibeam and sidescan sonar systems. He has participated in more than thirty-five research expeditions at sea since 1994, serving four times as a chief scientist and three times as a co-chief scientist. Brian is also a certified electrician, as well as a PADI- and AAUS-certified divemaster, rescue diver, and medic. He holds a bachelor's degree in marine science from Coastal Carolina University and a master's degree in marine science from USF.

David F. Naar, Ph.D., has a broad-based background in marine geology and geophysics. He earned a bachelor of arts degree in geology from UCSB in 1978 with an emphasis in geophysics. In 1990 he received a doctoral degree in earth sciences from the Scripps Institution of Oceanography. In 1989 David accepted a position as an assistant professor at the University of South Florida (USF) in the College of Marine Science (formerly a department of the College of Arts and Sciences). Since that time, he has remained active in mid-ocean research with an emphasis on microplate tectonics and hotspot–ridge interactions in the South Pacific. He has joined British, French, Spanish, Chinese, Japanese, Canadian, Swiss, and German investigators on several international multibeam mapping expeditions in the South Pacific, including a French submersible investigation of an active volcano aboard IFREMER's deep submersible, *Nautile*. In the spring of 1999, David obtained a Kongsberg Simrad EM-3000 multibeam system through an Office of Naval Research (ONR) Defense University Research Instrumentation Program award, which has been used to support ONR's HyCODE and Mine Burial and Scour studies. This system has been pooled with sidescan and seismic systems under a new Center for Coastal Ocean Mapping at USF, which David codirects. He has served as an associate editor for the *Journal of Geophysical Research and Marine Geophysical Researches,* and has co-authored more than forty publications. David has been at sea during fifty expeditions over the past twenty-two years, including several legs as chief and cochief. His current mapping efforts include shallow coastal areas in American Samoa, the Bahamas, and Florida, funded by the Office of Naval Research, the National Oceanic and Atmospheric Administration, the National Science Foundation, and the U.S. Geological Survey.

CONTACT THE AUTHORS

Dawn J. Wright
Department of Geosciences
104 Wilkinson Hall
Oregon State University
Corvallis, OR 97331-5506
Telephone: (541) 737-1229
Fax: (541) 737-1200
dawn@dusk.geo.orst.edu
dusk.geo.orst.edu

Brian Donahue
Telephone: (727) 553-1121
donahue@marine.usf.edu

David F. Naar, Co-Director
Telephone: (727) 553-1637
naar@usf.edu

Center for Coastal Ocean Mapping
College of Marine Science
University of South Florida
140 Seventh Avenue South
St. Petersburg, FL 33701-5016
Fax: (727) 553-1189
moontan.marine.usf.edu

Chapter 4

Using GIS to Track Right Whales and Bluefin Tuna in the Atlantic Ocean

ROB SCHICK

EDGERTON RESEARCH LAB

NEW ENGLAND AQUARIUM

BOSTON, MASSACHUSETTS

ABSTRACT ▶ While interdisciplinary research efforts have examined the distribution of pelagic species in the ocean, seldom has a synoptic link been made between these species and their oceanic environment. This link—making possible spatially accurate decisions concerning both species and environment—can be vital to the survival of threatened species. For example, the northern right whale remains one of the most critically endangered large whales in the world. While general patterns about their life history are known, better, more spatially explicit research efforts need to be made to ensure protection of the right whale. GIS, because it can link spatial data from different sources and in differing formats, is integral to these efforts.

Because nearly all our data has some spatial component, using GIS to store and analyze it greatly aids us in our quest to better understand endangered pelagic species with (in some cases) oceanwide distribution patterns. To that end, several different observation platforms have been put in use. Our research has included ship-, aerial-, and satellite-based platforms for both bluefin tuna and right whales. Satellite tagging efforts have also been used to identify movements of these species independent of research and fishing operations. These new approaches have led to remarkable breakthroughs in our understanding of the migration and behavior of right whales and bluefin tuna.

In the end, scientists are interested not only in where whales and pelagic fish go, but also what the environment is like at that location. Studying the ocean is an extremely complex endeavor, and differs from terrestrial studies in that the habitat has four dimensions. There are the x- and y-dimensions of the sea surface, but also the z-dimension of ocean depth. This z-dimension is important and dynamic, as environmental changes in the z-dimension of the ocean are often much more pronounced than in the x- and y-dimensions. Finally, there is a dimension of time. While large-scale patterns in the ocean may seem stable over time, there is a great deal of small-scale variation that is just now being integrated and understood. As part of our research efforts to understand the link between endangered species and their oceanic environment, we have used GIS to plot distribution and abundance patterns, as well as patterns in right whale and bluefin tuna distribution and movement in the Gulf of Maine and the western North Atlantic.

INTRODUCTION

Creating GIS coverages of species dynamics, their geographic distribution, and the associated environmental variables has proved a crucial first step. Visualizing the data in this way allows researchers to formalize original research questions, and, perhaps more importantly, to explore new research questions and directions. The goals of the GIS-based research at the New England Aquarium (NEAq) include: (1) documenting the habitat use patterns of right whales and bluefin tuna; (2) synthesizing data from a variety of research platforms; and (3) using spatial analysis to quantify the species–environment relationship.

RESEARCH

RIGHT WHALES

Only an estimated three hundred right whales (*Eubalaena glacialis*) survive in the North Atlantic (figure 1), despite more than sixty-five years of protection from commercial whaling. Although protected in the United States under the Endangered Species Act and the Marine Mammal Protection Act, the species has shown little sign of recovery. Population growth has been compromised by low reproductive rates (Knowlton et al. 1994) and high levels of human-caused mortalities (Kraus 1990; Kenney and Kraus 1993). These factors suggest an estimate of fewer than two hundred years to extinction (Caswell et al. 1999).

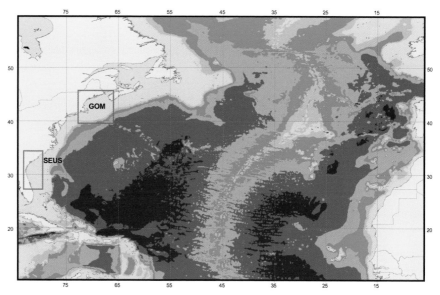

Figure 1. Map of the North Atlantic Ocean depicted using the ETOPO-5 elevation database from the USGS. Red boxes highlight two of the major research areas for NEAq: the southeastern United States and the Gulf of Maine. Research on the right whale takes place in each area, while research on bluefin tuna takes place in the Gulf of Maine, and via the pop-up archival tags, over the entire North Atlantic.

In the western North Atlantic, right whales occur in five areas along the east coast of North America, although at least two other undiscovered areas are suspected. The primary calving ground is found in the coastal waters of the southeastern United States during the winter months (figure 2). In the spring, aggregations of right whales are observed in the Great South Channel, east of Cape Cod, and in Massachusetts Bay (Schevill et al. 1986; Winn et al. 1986). In the summer and fall, right whales are observed in the Bay of Fundy, between Maine and Nova Scotia, and in an area on the Nova Scotian Continental Shelf, 50 kilometers south of Nova Scotia (Kraus et al. 1982; Stone et al. 1988) (figure 3).

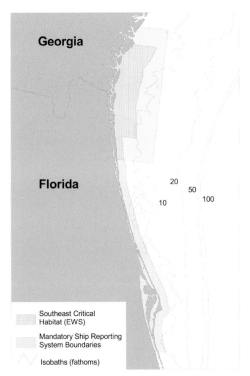

Figure 2. Detail of the southeastern United States study area, with the Right Whale Conservation area and isobaths drawn.

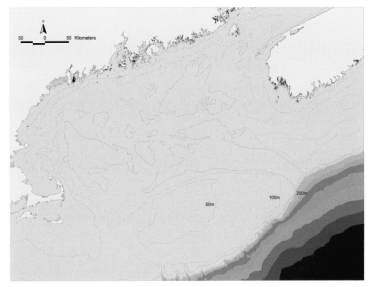

Figure 3. Detail of the Gulf of Maine study area, with 50-, 100-, and 200-meter depth contours drawn.

Researchers from the NEAq, the Center for Coastal Studies, National Marine Fisheries Service (NMFS), University of Rhode Island (URI), and many other collaborators collect data on the endangered right whale from a variety of platforms, including planes, research vessels, and satellite tags mounted on whales in many of the above habitats. The satellite data is particularly compelling to work with, because it offers a unique and unbiased look into movement patterns of right whales. Too often, researchers look where they think whales are going to be. This can be detrimental to statistical analysis, but is a necessary artifact of working with a free-ranging pelagic species that spends most of its time underwater. Thus, creating maps showing reconstructed tracks of satellite-tagged whales enables researchers to examine where whales go when they are not being actively followed by humans. The NEAq tagged several whales in 1996 and 1997 (Slay and Kraus 1998; Slay et al. in prep.), and from those efforts, the distribution and movement for three whales in the Gulf of Maine (GoM) are being examined.

One of these whales, number 1125, is of special interest because its track looks as if there are distinct behavioral changes along its length, which may reflect the whale's response to certain foraging areas (figure 4). Whale 2135 has a less detailed track than 1125 (figure 5), but is relevant because it tracks along the north edge of the Georges Bank region, an important habitat for right whales in the GoM. Finally, whale 1812, a mother and calf pair tracked by NEAq researchers, provided a great deal of life-history information as her movements tied together several important phases of the annual migration cycle for right whales (figure 6). She was initially tagged in the southeastern United States. After six weeks, she migrated with her calf north along the eastern coast of the United States into the GoM, where she remained for the duration of the tag. The satellite track of 1812 remains the longest track of a large whale in the Atlantic Ocean.

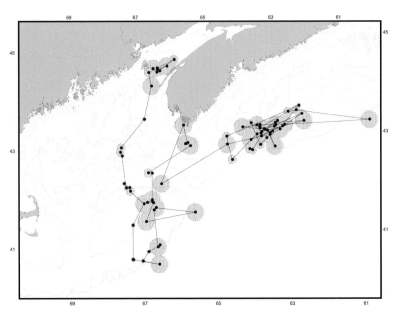

Figure 4. Reconstruction of the movement track of whale 1125, a mature female, as recorded by the ARGOS satellite system. Because there is positional error in each of the satellite fixes, a lookup table was created in ArcInfo to buffer the positions according to their accuracy. While the black dot represents the fix returned by the satellite, the polygon indicates that the animal could have been anywhere within that radius.

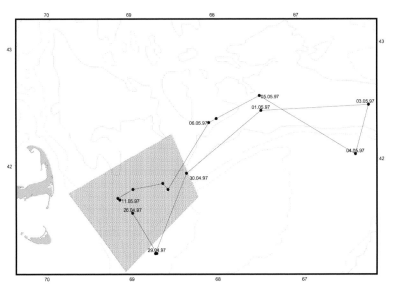

Figure 5. Reconstructed track of whale 2135, a male right whale tagged in the Great South Channel region of the GoM. This whale moved through an area of both high right whale density and high shipping traffic.

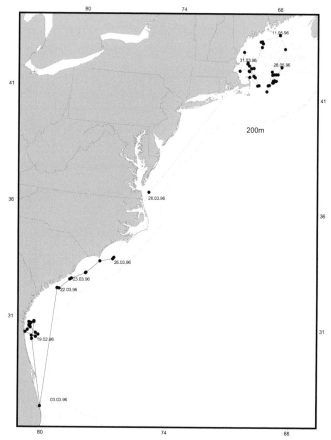

Figure 6. Reconstructed track of whale 1812, a mother and calf pair that migrated up the length of the east coast and into the GoM. Dotted lines indicate where the connection between two satellite fixes crossed land.

BLUEFIN TUNA

Relatively little is known about bluefin tuna (*Thunnus thynnus*) residency patterns in the GoM (Lutcavage et al. 2000), and even less is known about the habitats the fish occupy once they leave the GoM. Bluefin tuna are found in the GoM from late May through November, and are usually found in schools of tens or hundreds of individuals (Crane 1936; Bigelow and Schroeder 1953; Mather 1962). Yet schools with more than five thousand surface fish have been noted (Lutcavage and Kraus 1995; Lutcavage et al. 1997). The bluefin tuna arrive to feed on seasonal aggregations of prey species such as herring mackerel and sand lance (Crane 1936). After feeding in the GoM, it has long been thought that bluefin return to the Gulf of

Mexico to spawn (Richards 1976), yet recent studies conducted by NEAq researchers have shown that once bluefin leave the GoM, they range over the entire northwest Atlantic (Lutcavage et al. 1999).

To better understand these multiscale distribution patterns, the NEAq's research has several components. The first goal is to characterize the abundance and behavior of surface schools of bluefin tuna in the GoM (figure 7). The second goal is to document where the fish go when they leave the GoM, and what ecological factors drive their impressive transatlantic migrations (figure 8). A crucial third goal is to incorporate the distribution information with oceanographic data to better understand the links between distribution, behavior, abundance, and the ocean environment. This effort uses GIS and spatial statistics to analyze the distribution patterns of tuna schools in the GoM. Initially, the focus is on the distribution and abundance of tuna schools in relation to ocean structure and productivity.

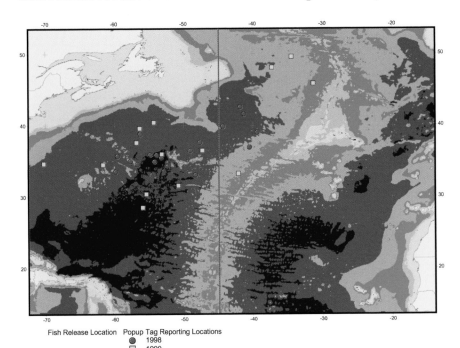

Fish Release Location   Popup Tag Reporting Locations
● 1998
□ 1999

Figure 7. Reporting locations for giant bluefin tuna tagged in the GoM in 1997 and 1998. Also shown in red is the 45-degree meridian, which is the international stock boundary line for bluefin tuna. Bluefin tuna are managed as separate eastern and western populations and are assumed to have low mixing rates.

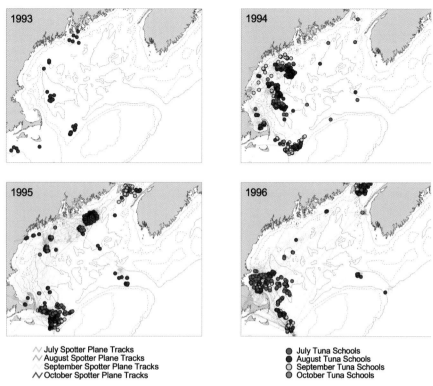

Figure 8. Results from four consecutive years of aerial survey research on the distribution patterns of bluefin in the GoM. During 1994–1996 the planes were equipped with GPS units and laptop computers to serve as data loggers. This equipment enabled NEAq researchers to keep track of where the planes flew on a given day. For some days, equipment error precluded continuous recording of the plane's track.

OCEANIC VARIABLES

One data set of special interest to researchers here at the NEAq is the sea surface temperature (SST) fronts data set processed and stored at the URI's Graduate School of Oceanography (Ullman and Cornillon 1999). The importance of fronts in the distribution patterns of pelagic species has been noted for many years (Nakamura 1965; Olson et al. 1994), and the quantitative relationship between fronts and pelagic fish has been investigated for swordfish (Podestá et al. 1993) and for albacore and skipjack tuna (Laurs et al. 1984; Fiedler and Bernard 1987). An SST front simply represents a temperature discontinuity in the ocean, as observed by orbiting satellites (figure 9). These temperature discontinuities are located using an edge detection algorithm, and each located front has data associated with it, such as the strength of the front, the temperature along the front, and the mean direction to warm water (Ullman and Cornillon 1999). This data allows NEAq researchers to investigate and quantify the spatial relationship between foraging tuna and different types of fronts.

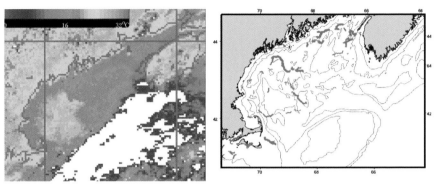

Figure 9. The left panel is a sea surface temperature image from August 8, 1994, and the right panel shows a line coverage of the SST fronts as interpolated from the above image. The magnitude of the temperature gradient across the front is reflected by the thickness of the lines. Note that the large white region in the top panel represents clouds, thus there are no fronts detected for that area.

The Distributed Oceanographic Data System (DODS), an online data distribution system (unidata.ucar.edu/packages/dods) developed by Peter Cornillon's lab at URI, provides a wealth of oceanographic data. With the help of a colleague (Paul Goulart), the author developed a script that took an ASCII text list of days for which data needed to be retrieved, and then used the DODS system to retrieve data over the Web, parse it into a data structure, and write out ASCII text files with both location and attribute information (figure 10). The script was developed using ESRI® Arc Macro Language (AML) software to take the raw data, generate line coverages, add attribute data to each line, create interpolated grids of Euclidean distance to each front, and ascribe projection information to each coverage. This AML allowed for the creation of a set of environmental coverages designed to examine the daily effects of fronts in the GoM, but researchers have also proposed that fronts may act as mechanisms that aggregate passive dispersing prey species (Olson et al. 1994; Podestá et al. 1993). If true, this aggregation may take time to develop. To investigate the possibility, a second AML was created that examined the fronts in a two-week period. From this data, interpolated grids of frontal density can be created. The two different types of grids allow for the investigation of different types of biological

questions: the first for examination of daily interactions with sometimes ephemeral fronts, and the second for examination of distribution in relation to persistent fronts.

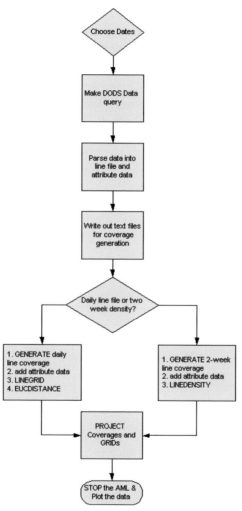

Figure 10. Flow chart representing the flow of SST front data from the DODS system to
ArcInfo coverages and grids.

SYNTHESIS AND ANALYSIS

Right whales are faced with a real extinction threat (Caswell et al.
1999). In light of this risk, it is important to consider the spatial and
temporal aspects of current conservation measures as they relate to
right whale distribution. Right whales face mortality threats from

entanglement in fishing gear, from ship strikes, and potentially from sources such as pollution and habitat loss. Numerous conservation measures have been put in place in the GoM as well as the southeastern United States to protect the right whale. These measures include the early warning system (EWS) of aerial surveys in the southeastern United States, as well as the Mandatory Shipping Reporting System (National Oceanic and Atmospheric Administration, NMFS Office of Protected Resources). GIS aids in examining both the spatial and temporal overlap between these conservation measures and right whale distribution patterns. For example, whale 1812, along with her calf, was seen in the southeastern United States during the time of the EWS (figure 11a), migrated north along the coast, and then was seen again in the GoM (figure 11b). When whale 1812 entered the GoM, it traveled through an area of intense shipping traffic and recrossed the shipping lane into Boston Harbor (figure 11b). Research efforts are underway here at the NEAq and elsewhere (Florida Marine Research Institute) to quantify the interaction between right whales and anthropogenic activities.

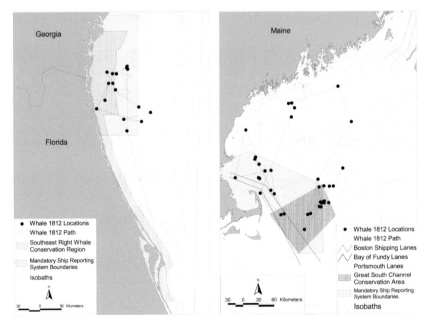

Figure 11. Reconstructed portions of whale 1812's track in (a) the southeastern United States, and (b) the GoM. GIS can be used to look at the spatial overlap between whale movements and conservation areas (a), as well as areas of high shipping traffic (b). The yellow area in (b) is a new system initiated by the International Maritime Organization and the National Marine Fisheries Service, which asks ships to notify the port of call upon entering this area.

Bluefin tuna are found in the GoM during the summer and fall months. While the timing of their migration into the GoM as well as their distribution and abundance patterns can vary from year to year, it is generally assumed that bluefin are in the GoM to feed (Crane 1936). Thus the future research objectives for the aerial survey database (Lutcavage et al. 1997) are to examine the spatial relation between bluefin and environmental parameters associated with foraging and aggregation (Lutcavage and Goldstein in prep.). Previous research on large pelagic fish species has noted the importance of SST fronts (Laurs et al. 1984; Fiedler and Bernard 1987; Podestá et al. 1993), thus researchers are examining bluefin distribution and abundance with reference to ephemeral (e.g., daily) fronts, as well as the density of persistent fronts. The relation between bluefin tuna distribution and the SST fronts in late July 1994 is striking (figure 12). GIS allows for the examination of both the spatial correlation and the correlation between different behaviors and different types of fronts. Each of the fronts has attribute data associated with it, such as the strength of the temperature gradient, and the direction to warm water. Within this GIS framework one can query the different types of relationships in order to learn more about bluefin foraging behavior in the GoM.

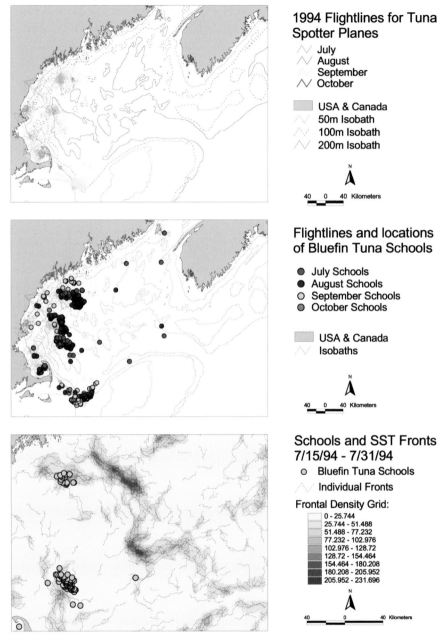

Figure 12. Tracks of spotter planes in the GoM: (a) used to document locations of bluefin tuna schools, and (b) the spatial overlap between tuna schools and areas of high SST frontal density in one two-week period. It is interesting to note that the documented tuna schools in this timeframe were all seen in areas of comparatively high frontal density.

A quantitative spatial analysis of the species–environment relationship for bluefin is necessary because each of these variables, whether species or environmental, has a great deal of spatial structure and autocorrelation. To examine this spatial relationship, the family of simple and partial Mantel tests is used. These tests are essentially correlation analyses of similarity between two or more matrices (Mantel 1967; Smouse et al. 1986). For example, these techniques can be used to look at the similarity between species distribution and environmental patterns while accounting for spatial autocorrelation (Legendre 1993; Legendre and Fortin 1989; Leduc et al. 1992; Schick and Urban 2000). The test differs from a typical correlation analysis or a multiple regression framework in that the values being tested are degrees of dissimilarity or distance between locations. The power of the test lies in its ability to partial out effects of confounding variables to enable researchers to identify pure spatial correlations.

In research at the NEAq, the Mantel test is used in a variety of ways, including examining the differences between the movement patterns of right whales as they correlate to the environment, and conducting population level analysis of the distribution of bluefin tuna in relation to ocean features such as SST fronts. For the bluefin tuna, several different matrices were created, including one "species" matrix of sightings per unit effort of fish in the GoM, spatial separation between tuna schools, and numerous environmental matrices of depth, slope, temperature, distance to a front, and density of persistent fronts. Testing all of these in a Mantel framework enables the examination of the spatial relationship of bluefin and frontal features in the GoM with high spatial and temporal resolution.

FUTURE STUDIES

From both a science and data management perspective, one of the more difficult challenges we have encountered is finding a way to integrate the two-dimensional data of satellite tracks with the three-dimensional patterns in the ocean. For some of the variables used in NEAq research, notably SST fronts, accounting for the depth dimension is unnecessary because the frontal data is derived from a 2-D image of the ocean's surface. Yet, with this data the time dimension becomes relevant, since there can be more than four AVHRR (Advanced Very High Resolution Radiometer) images in any given day.

2-D species distribution patterns, however, must be interpreted in the appropriate context. For NEAq research, examining the relationship between surface patterns of these species and surface patterns in the ocean is valid, yet in this analysis we must acknowledge that whales, for example, spend more than 90 percent of their time underwater, and surface fronts are simply that: surface expressions of complicated fluid processes taking place at depth. GIS can be used to plot the 2-D movement patterns of a tagged fish (figure 13a), but this can be misleading if not interpreted in three dimensions. For example, this track actually had depth information along with x- and y-coordinates, and visualizing this track using the ArcView 3D Analyst™ extension gives an idea of what the track actually looked like in the ocean (figure 13b).

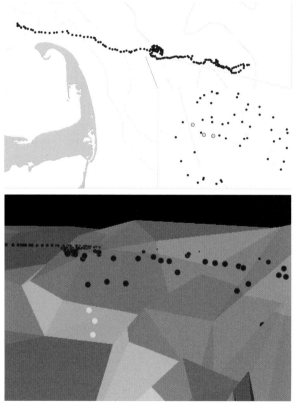

Figure 13. Example of the different information conveyed between a two-dimensional (2-D) map and a three-dimensional (3-D) image of a tagged bluefin tuna in the western portion of the Great South Channel in the GoM. The point of view in the second panel is from the northwest, looking south at the track. The three selected points (in yellow) in the 2-D close-up are the same three points in the 3-D image. The second panel gives an idea of the true path in 3-D.

Therefore, researchers must keep looking forward to future developments in three- and four-dimensional GIS. Since each of these target species spends most of its time underwater, a challenge for biological oceanographers and marine ecologists is to find ways to (1) track these species as they move throughout the ocean depths, and (2) to find ways to correlate these movements and distribution patterns with changes in the ocean habitat. Both ArcView GIS and ArcInfo have some three-dimensional capabilities, yet neither system can incorporate animal movements and relate them, for example, to changes within the water column. Future GIS developments should attempt to solve the problem of relating three-dimensional movements of animals to changes in the fluid environment of the ocean, leading to an enhanced understanding of the relationship between pelagic species and their ocean environment.

ACKNOWLEDGMENTS

The author would like to thank Paul Goulart, who helped write the code necessary to access and format the fronts data from DODS. Thanks are also extended to Peter Cornillon and Dan Holloway from URI for all of their help and patience in mastering DODS. Charles Convis and the ESRI Conservation Program provided the author with a grant to take an AML course. Erika Shore, Scott Kraus, and Molly Lutcavage provided helpful comments on earlier versions of this chapter. This work was supported by an Office of Naval Research grant to Molly Lutcavage and Scott Kraus.

REFERENCES

Bigelow, H. B., and W. C. Schroeder. 1953. Fishes of the Gulf of Maine. *Fisheries Bulletin US* 53:1–577.

Caswell, H., M. Fujiwara, and S. Brault. 1999. Declining survival probability threatens the North Atlantic right whale. *Proceedings of the National Academy of Sciences* 96:3,308–13.

Crane, J. 1936. Notes on the biology and ecology of giant tuna, *Thunnus thynnus Linneaus,* observed at Portland, Maine. *Zoologica* (New York Zoological Society) 21:207–11.

Fiedler, P. C., and H. J. Bernard. 1987. Tuna aggregation and feeding near fronts observed in satellite imagery. *Continental Shelf Research* 7(8):871–81.

Kenney, R. D., and S. D. Kraus. 1993. Right whale mortality—a correction and an update. *Marine Mammal Science* 9(4):445–46.

Knowlton, A. R., S. D. Kraus, and R. D. Kenney. 1994. Reproduction in North Atlantic right whales. *Canadian Journal of Zoology* 72:1,297–305.

Knowlton, A. R., P. K. Hamilton, M. K Marx, S. M. Martin, and S. D. Kraus. 1999. Maintenance of the right whale catalog 1 January–31 December, 1999. *Contract Report for National Oceanic and Atmospheric Administration, Northeast Fisheries Science Center.*

Kraus, S. D., J. H. Prescott, P. V. Turnbull, and R. R. Reeves. 1982. Preliminary notes on the occurrence of the North Atlantic right whale (*Eubalaena glacialis*), in the Bay of Fundy. *Reports of the International Whaling Commission* 32:407–11.

Kraus, S. D., K. E. Moore, C. E. Price, M. J. Crone, W. A. Watkins, H. E. Winn, and J. H. Prescott. 1986. The use of photographs to identify individual North Atlantic right whales (*Eubalaena glacialis*). *Reports of the International Whaling Commission,* Special Issue 10:139–44.

Kraus, S. D., M. J. Crone, and A. R. Knowlton. 1988. The North Atlantic right whale. In *Audubon Wildlife Report 1988/1989,* W. J. Chandler, ed. New York: Academic Press.

Kraus, S. D. 1990. Rates and potential causes of mortality in North Atlantic right whales (*Eubalaena glacialis*). *Marine Mammal Science* 6(4):278–91.

Laurs, R. M., and R. J. Lynn. 1977. Seasonal migration of North Pacific albacore, *Thunnus alalunga,* into North American coastal waters: distribution, relative abundance, and association with Transition Zone waters. *Fisheries Bulletin* 75:795–822.

Laurs, R. M., P. C. Fiedler, and D. R. Montgomery. 1984. Albacore tuna catch distributions relative to environmental features observed from satellites. *Deep-Sea Research* 31(9):1,085–99.

Leduc, A., P. Drapeau, Y. Bergeron, and P. Legendre. 1992. Study of spatial components of forest cover using partial Mantel's tests and path analysis. *Journal of Vegetation Science* 3:69–78.

Legendre, P. 1993. Spatial autocorrelation: trouble or new paradigm? *Ecology* 74:1,659–73.

Lutcavage, M., R. Brill, G. Skomal, B. Chase, and P. Howey. 1999. Results of pop-up satellite tagging on spawning size class fish in the Gulf of Maine. Do North Atlantic bluefin tuna spawn in the Mid-Atlantic? *Canadian Journal of Fisheries and Aquatic Science* 56:173–77.

Lutcavage, M., J. Goldstein, and S. Kraus. 1997. Distribution, relative abundance, and behavior of giant bluefin tuna in New England waters, 1995. *International Commission on Conservation of Atlantic Tunas Collective Volume of Scientific Papers,* Madrid, Spain, SCRS/96/129.

Legendre, P., and M-J. Fortin. 1989. Spatial pattern and ecological analysis. *Vegetation,* 80:107–38.

Mantel, N. 1967. The detection of disease clustering and a generalized regression approach. *Cancer Research* 27:209–20.

Mather, F. J. 1962. Tunas (genus *Thunnus*) of the western North Atlantic. Part III. Distribution and behavior of *Thunnus* species. In *Proceedings of the Symposium on Scombroid Fishes. Part I,* Marine Biological Association of India, Mandapam Camp, 1–6.

Nakamura, H. 1969. *Tuna Distribution and Migration.* London: Fishing News (Books) Ltd.

Olson, D. B., G. L. Hitchcock, A. J. Mariano, C. J. Ashjian, G. Peng, R. W. Nero, and G. P. Podestá. Life on the edge: Marine life and fronts. *Oceanography* 7(2):52–60.

Podestá, G. P., J. A. Browder, and J. J. Hoey. 1993. Exploring the association between swordfish catch and thermal fronts on the U.S. longline grounds in the western North Atlantic. *Continental Shelf Research* 13:252–77.

Richards, W. J. 1976. Spawning of bluefin tuna (*Thunnus thynnus*) in the Atlantic Ocean and adjacent seas. *International Commission on Conservation of Atlantic Tunas Collective Volume of Scientific Papers* 5(2):267–78.

Schevill, W. E., W. A. Watkins, and K. E. Moore. 1986. Status of right whales (*Eubalaena glacialis*) in Cape Cod Waters. *Fishery Bulletin* 80(4):875–80.

Schick, R. S., and D. L. Urban. 2000. Spatial components of bowhead whale (*Balaena mysticetus*) distribution in the Alaskan Beaufort Sea. *Canadian Journal of Fisheries and Aquatic Science* 57(11):2,193–200.

Smouse, P. E., J. C. Long, and R. R. Sokal. 1986. Multiple regression and correlation extensions of the Mantel test of matrix correspondence. *Systematic Zoology* 35:627–32.

Stone, G. S., S. D. Kraus, J. H. Prescott, and K. W. Hazard. 1988. Significant aggregations of the endangered right whale, *Eubalaena glacialis*, on the continental shelf of Nova Scotia. *Canadian Field Naturalist* 102:471–74.

Ullman, D. S., and P. C. Cornillon. 1999. Satellite-derived sea surface temperature fronts on the continental shelf off the northeast U.S. coast. *Journal of Geophysical Research* 104:23, 459–78.

Winn, H. E., C. A. Price, and P. W. Sorensen. 1986. The distributional biology of the right whale (*Eubalaena glacialis*) in the western North Atlantic. *Reports of the International Whaling Commission* Special Issue 10:129–38.

ABOUT THE AUTHOR

Rob Schick is an ecologist at the New England Aquarium's Edgerton Research Lab. His professional interests include using GIS and spatial analysis to better understand the species–environment relationship in the marine realm. Rob has nearly finished a master's degree in landscape ecology from Duke University, and is interested in exploring the extent to which principles and tools from landscape ecology can be applied to the marine environment. His master's research focused on the use of spatial statistics to analyze bowhead whale distribution relative to oil exploration activities in the Alaskan Beaufort Sea. Prior to studying at Duke, Rob completed a B.S. in zoology at the University of Washington. Rob lives outside Boston with his wife and their two dogs.

CONTACT THE AUTHOR

Rob Schick
Edgerton Research Lab
New England Aquarium
Central Wharf
Boston, MA 02110-3399
Telephone: (617) 973-5200
Fax: (617) 723-9705
rschick@neaq.org
www.marinegis.org

Chapter 5 — Finding the Green Under the Sea: The Rehoboth Bay Ulva Identification Project

KIMBERLY B. COLE, DAVID B. CARTER

DELAWARE COASTAL PROGRAMS

DEPARTMENT OF NATURAL RESOURCES AND ENVIRONMENTAL CONTROL

DOVER, DELAWARE

CHUCK SCHONDER

SEA DOG SOLUTIONS

SAN DIEGO, CALIFORNIA

MARK FINKBEINER AND RENÉE E. SEAMAN

NOAA COASTAL SERVICES CENTER

CHARLESTON, SOUTH CAROLINA

ABSTRACT Every year more boaters, fishing enthusiasts, and other recreational users place a greater strain on the limited space and natural resources of Rehoboth Bay. Regional development and agricultural and industrial activities affect the bay as well, with increased runoff of chemicals and nutrients. A particular genus of macroalgae known as *Ulva* grows in these waters naturally, but has become an increasing problem as the chemistry of the bay changes. *Ulva* management in Rehoboth Bay has challenged government agencies and local organizations for a number of years. Increases in *Ulva* biomass can lead to an increase in dead organic matter and to reduced dissolved oxygen levels in the bay. Lower dissolved oxygen can have a significant impact on the local aquatic ecology. Increased dead organic loads can cause foul odors and create hazards for boats.

Government agencies have funded *Ulva* collection operations to reduce the effects of the macroalgae on the ecosystem and recreational opportunities in Rehoboth Bay. Unfortunately, little data exists on either the spatial extent of *Ulva* in Rehoboth Bay or the effectiveness of the collection operations. It seems clear that the agencies funding these operations must identify the present extent of *Ulva* in the bay and periodically update that information to determine the effectiveness of their efforts.

This chapter describes how the Delaware Coastal Programs (DCP) identified the spatial extent of macroalgae, including *Ulva*, in Rehoboth Bay in the late spring of 1999 and 2000. DCP used aerial photography, image-processing software, a geographic information system (GIS), and a limited field survey to do this work, which resulted in the identification of nearly 2 square kilometers of macroalgae in all but the deepest parts of the bay. The project results illustrate the significant potential benefits that remotely sensed imagery brings to resource management work.

INTRODUCTION

During the spring of 1999, we identified the spatial extent of macroalgae in the shallow portions of Rehoboth Bay using traditional photogrammetric methods. We used true color aerial photographs, image-processing software, a GIS, and a limited field survey to identify 1.88 square kilometers of macroalgae in all but the deepest parts of the bay. Turbid conditions prevented identification and measurement of the full extent of the vegetation.

Although the 1999 effort was successful, it was clear that aerial photography could penetrate neither the deeper parts of the bay, nor where conditions were turbid. For another effort in 2000, the DCP joined forces with the National Oceanic and Atmospheric Administration (NOAA) Coastal Services Center benthic-mapping project. This project, a part of the center's Coastal Remote Sensing program, uses a RoxAnn acoustic sensor to identify benthic cover in turbid areas up to 40 meters deep. This instrument classifies bottom type by extracting data on bottom roughness and bottom hardness from the primary and secondary sounder echoes. The unit is able to collect data throughout the bay in areas greater than 1.4 meters in depth, and complements the aerial photography well. RoxAnn data is easily exported into a GIS for postprocessing and integration with other spatial data for the bay. The result will be a comprehensive distribution map of *Ulva* within Rehoboth Bay.

PROJECT PLANNING

It has been well established that aerial photography can be used to identify submerged aquatic vegetation (SAV), given the proper environmental conditions. Regular color aerial photography has been used to penetrate water in numerous projects along the U.S. east coast and in the Caribbean well beyond the maximum depths of Rehoboth Bay. We referred to such projects to determine the feasibility of mapping the macroalgae *Ulva* in Rehoboth Bay.

Working closely with an aerial photography firm, Keystone Aerial Surveys, Inc., we began in March 1999 to find the best way to take the photographs. Four north–south flight lines were agreed upon at a scale of 1:12,000, with 60 percent overlap on the ends and 30 percent overlap on the sides (figure 1). To relate the resulting images to other spatial data stored in our and the Department of Natural Resources and Environmental Control's (DNREC) databases, the images needed to conform to the coordinate system used by the other spatial data. This operation required known ground control points (GCP) in each image. Because some images covered only or mostly water, they would not have easily identifiable GCPs. On May 17, 1999, we anchored eight white buoys, each with a 16-square-foot surface area, in the bay to act as GCPs. Division of Soil and Water Conservation staff built and provided the buoys. While deploying the buoys, we recorded their locations with differential Global Positioning System (DGPS) equipment.

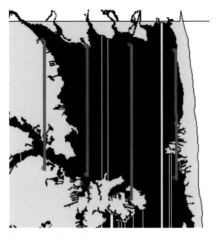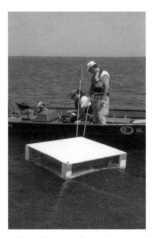

Figure 1. Flight lines and ground control points utilized for the first photographic mission in 1999 and repeated in 2000.

ENVIRONMENTAL CONDITIONS

Environmental conditions such as tide, turbidity, sun angle, weather, and platform altitude can all affect the penetration depth of an aerial photo. To allow the deepest possible penetration of water and clearest identification of the macroalgae, we chose to capture the images using a midmorning flight within two hours of low tide, on a sunny day, with low wind speeds, prior to the opening of the tourism season (May 31, 1999). The low wind speed would decrease sun scatter from small waves. A morning flight before May 31 would decrease boat traffic and thus decrease turbidity. Low tide would allow the deepest absolute penetration with a given set of conditions. We and Keystone staff identified a number of possible days and times for the flight. The conditions on the first possible day, May 22, met our environmental condition requirements, and Keystone took the photographs on that day. After the flight, on May 26, we relocated the buoys using GPS to identify possible movement of the GCPs for coordinate registration purposes. The buoys were also removed from the bay on that day.

PREPROCESSING PRODUCTS

On June 11, 1999, Keystone delivered twenty-eight 9-by-9-inch black-and-white photos covering Rehoboth Bay, the surrounding land, and the adjacent Atlantic coastline (figure 2).

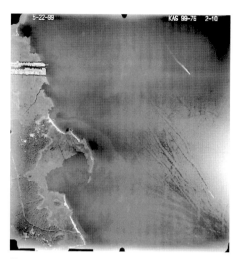

Figure 2. Keystone Aerial Surveys, Inc., flew four north–south transects along Rehoboth Bay on May 22, 1999. DCP received the images in two formats: 9-by-9-inch black-and-white hard-copy photos, as seen above, and real-color scanned versions.

The photos represented the bay at a scale of 1:12,000 as our contract specified. Fourteen compact disks with real-color electronic images of the same area were included in the package. The 1,000-dots-per-inch scanning process used to create the electronic files resulted in a 0.3-meter resolution for the color images. Although the large file sizes (approximately 300 megabytes per image) produced by the high-resolution images taxed our processing power, these images will provide an excellent base for later comparison studies or other detailed work in the Rehoboth Bay area.

DIGITAL AND SPATIAL PROCESSING

The first step in preparing aerial photos for analysis with other data involves registering the images to a coordinate system. We used ERDAS IMAGINE® image-processing software to register the images to the Delaware State Plane projection, North American Datum of 1983 (figure 3). The software used the GCPs mentioned earlier, as well as additional spatial reference files, to stretch and rotate the images to fit the chosen coordinate system. Because each file was so large, our fastest computer took more than two hours to process each image. This registration procedure nevertheless resulted in images that matched DNREC's base road files and other spatial data.

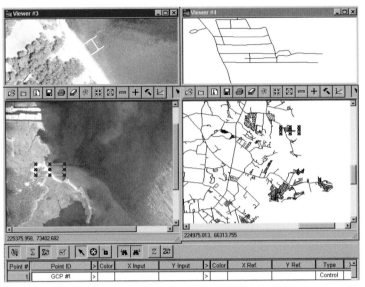

Figure 3. Example of ERDAS IMAGINE software registration of an image. This process was repeated for each of the twenty-eight images captured by Keystone.

After registering the images, DCP staff used ERDAS IMAGINE to make image subsets, clipping the images to remove residual photographic elements such as solar glare. This process resulted in much cleaner images for classification purposes. Once registered and clipped, the images were ready for classification. Image classification can be divided into supervised and unsupervised methods. In supervised methods, an analyst selects representative training sites of various habitats. These guide the statistical classification of the image data. In unsupervised methods, the computer classifies images based solely on statistical variations in image data. Both methods have documented positive and negative aspects (Wilkinson 1996). We used variations of both methods and compared the results (figure 4). In all cases, the resulting classifications included more omission of visible macroalgae and commission of other land and water types than we expected. Commission occurred when other land and water types were classified as macroalgae. After these initial attempts, we tried some higher-order image processing to improve the end results (figures 5 and 6).

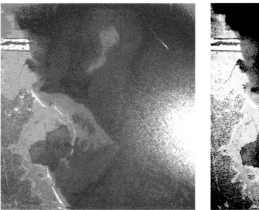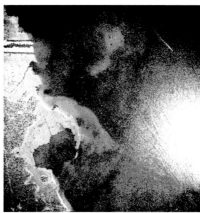

Figure 4. The supervised (left) and unsupervised (right) classifications identified scattered parts of visible macroalgae, but the gunshot-like results were deemed too inaccurate.

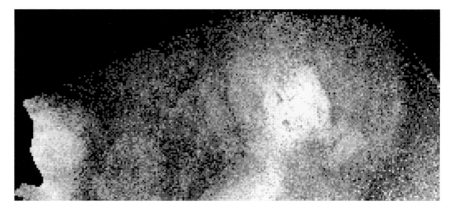

Figure 5. Example of an image that was processed using a land mask. Using ERDAS IMAGINE, the analysts cut out the land portions of the images and ran unsupervised classification algorithms on the water. This allowed the computer to distinguish more subtle differences in the water, but still resulted in unacceptable amounts of omission and commission.

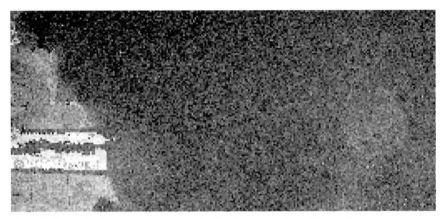

Figure 6. Example of an image processed with a high-pass filter. The analysts also tried a number of high-pass filters to emphasize changes in land and water types prior to classification. This step also did not result in significant classification improvements.

Although initial attempts at image processing suggested promising results given more time for refinement, they produced only partially adequate macroalgae classification (figure 7). Our analysts asked the NOAA Coastal Service Center staff for advice and guidance. Center staff explained that in a similar project in North Carolina, the most useful SAV classification came from visual interpretation hand digitized on a computer. We chose to do the same.

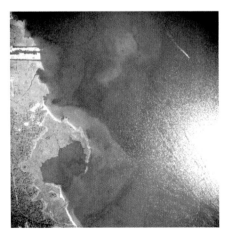 

Figure 7. From the registered and clipped images, the analysts visually identified macroalgae but were unable to distinguish between species. One macroalgae classification is represented in green.

Using ArcView GIS, the analysts identified and classified the macroalgae based on visual interpretation of the registered, clipped images. This resulted in the identification, mentioned above, of 1.88 square kilometers of macroalgae in all but the deepest parts of Rehoboth Bay. Although anecdotal evidence from buoy retrieval suggested macroalgae in the center of bay, the images could not penetrate deeply enough in that area because of the high turbidity.

A few weeks after Keystone took the photos, the Division of Water Resources had a number of DNREC staff carry out a limited survey of macroalgae species in Rehoboth Bay. Thirty-nine points along the bay shore were surveyed for the presence of *Ulva, Agardhiella,* and *Gracilaria.* The point data gathered in this survey could not be used in an assessment of the accuracy of the photo-derived data for a variety of reasons. A qualitative comparison, however, does indicate general agreement on the location of the macroalgae.

After classifying the images through visual interpretation, the individual files were still extremely large and difficult to manage. We used ERDAS IMAGINE to resample the images from a resolution under a meter to resolutions of 5 and 10 meters. This reduced the file sizes by 250 and 1,000 times, respectively. With smaller file sizes, our analysts were able to combine the individual images into a mosaic of the entire bay (figure 8).

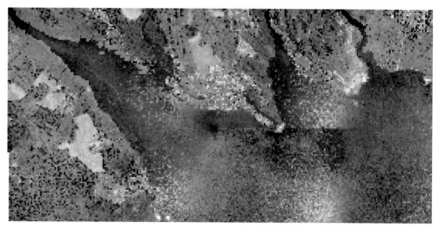

Figure 8. The mosaic of Rehoboth Bay created from the registered and clipped images. The green polygons signify the visually interpreted macroalgae. The work done with the 1999 aerial photos of Rehoboth Bay resulted in the classification of 1.88 square kilometers of macroalgae in all but the deepest parts of the bay.

This project revealed some of the challenges of working with high-resolution image data: the need, for example, for large amounts of storage space, a well-designed classification system, and the significant effort associated with image processing. Nevertheless, the project results from 1999 illustrate the large potential that remotely sensed imagery has in resource management work. Consequently, we decided to continue work in the bay through 2000 with more aerial photography and advanced remote sensing technology provided by the NOAA Coastal Services Center.

REMOTE SENSING WITH THE ROXANN ACOUSTIC SENSOR

Rehoboth Bay posed a unique remote sensing quandary: the bay was extremely shallow (depths did not exceed 7 feet at low-low tide) and naturally turbid. Aerial photogrammetry had proven unsuccessful for comprehensive mapping, and the bay was too shallow for multibeam or sidescan sonar. Staff at the NOAA Coastal Services Center's Coastal Remote Sensing program determined that Rehoboth Bay was an ideal study site for data collection using the RoxAnn acoustic sensor, which functions in water 1 to 30 meters deep and, when used in conjunction with ground-truthing, can provide detailed classification information.

Acoustic sensor surveys were carried out between June 12 and June 16, 2000. The survey was timed to occur as closely as possible to the date of the aerial photographic mission. Doing so helped to eliminate large differences in the biomass of algae as the summer progressed. Recreational boat traffic at that time was still relatively low as well. Variable field conditions prompted us to survey the bay using nontraditional grids. The bay was initially divided down the center north to south, and each section was successively divided. The bay was also surveyed east to west following the same method. With the variable tide and wind (5 to 15 knots) conditions prevailing during the week, this method provided the flexibility to maintain an equal coverage of bay; as time permitted, more transects were added. An estimated thirty survey lines (fifteen north–south, fifteen east–west), spaced approximately 350 meters apart, were completed, totaling more than 80,000 data points (figure 9).

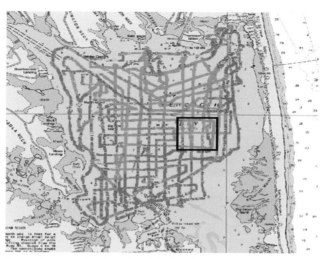

Figure 9. Navigational tracks of the RoxAnn survey color coded by bottom type: orange represents soft bare mud (less than 15 percent algae cover), yellow represents harder mud or sand (less than 15 percent algae), and green represents algae (greater than 15 percent algae). The blue box is the area extrapolated to create the 3-D bathymetric profile in figure 15.

The RoxAnn Groundmaster system consists of a signal processor built around a Furuno 6000 echosounder and a 20-kilohertz transducer mounted on the side of the boat approximately level with the keel (figure 10). The system was connected to a Garmin DGPS and a field laptop computer loaded with the RoxMap software (figure 11). Data was collected at an average speed of 5 knots per hour (9.2 kilometers per hour) in water deeper than 1 meter. Faster speeds and water shallower than 1.4 meters typically generated false depth data that was removed from the analysis during postprocessing.

Figure 10. Equipment used for the acoustic surveys, top to bottom: (a) a digital video camera in a waterproof housing is used to capture representative habitat images; (b) the RoxAnn Groundmaster system; (c) the hull-mounted transducer; and (d) the TOV-1 towed underwater camera.

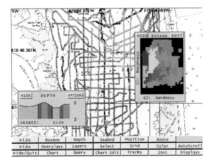

Figure 11. Sample of RoxMap software logging data in the field. Echoes are sensed by the transducer, transformed in the Groundmaster unit, and displayed in RoxMap. The software displays a depth profile and sea floor (as created by the user) in real time on a NOAA nautical chart.

Upon the detection of a unique data return, the Fishers towed underwater video camera (TOV-1) was deployed. This permitted observation of habitats that were obscured by turbidity. Five to ten minutes of Super VHS video were recorded at each station. Initial video interpretations were made on the boat and were used in the field calibration of the RoxAnn. Positional data was recorded at the start of each video, providing a reference for comparison during the final data analysis. A total of thirteen sites were observed with the TOV-1, and five were examined by snorkeling.

FIELD CLASSIFICATION

The RoxMap software was used to analyze and characterize the primary (roughness) and secondary (hardness) returns. Within RoxMap the acoustic returns were plotted on a graph in a window (the box) with the y-axis representing an index of roughness (E1), and the x-axis an index of hardness (E2; figure 12). Different bottom types create unique hardness and roughness signatures that form distinct clusters when plotted. Each cluster was then assigned a classification within the box. The classification was a user-specified (supervised) classification based on ground truthing with the TOV and by snorkeling. Changing the classification box does not alter data values; data could therefore be constantly collected, and classification developed on the fly. When a unique cluster was detected, the TOV was deployed and the clusters were characterized according to the video display.

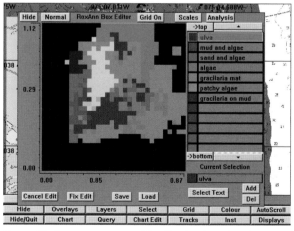

Figure 12. A screen capture of the RoxAnn box showing the relationships between bottom types based on roughness (y-axis) and hardness (x-axis).

RoxAnn records a point every second and records the following attributes: E1, E2, depth (meters), time, latitude, and longitude. The E1 and E2 values characterized a portion of the bottom (the footprint) that was one-tenth the depth, meaning that the data points recorded in the shallow bay were of a very small and very detailed area. The size of the footprint must be considered when trying to aggregate the data into meaningful classes. The following classes were developed for Rehoboth Bay: *Ulva, Gracilaria,* mixed algae, *Gracilaria* on mud, mud with 15 percent algae, sand with 15 percent algae, soft mud, mud, hard sand, sandy mud, bioturbated mud, shell/mud/algae, and unknown.

POSTPROCESSING

The classified point data generated in RoxMap was imported to an ArcView project (figures 13 and 14). Before analysis, erroneous or unreliable data points were identified and removed. These unreliable data points occur when the sensor is sounding in extremely shallow water (less than 1 meter) or if the transducer is unable to detect the primary or secondary echoes due to boat speed. False bathymetry values and repeating points indicate unreliable data.

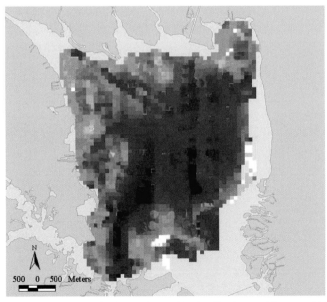

Figure 13. Bathymetric data collected by the RoxAnn and imported into ArcView, not corrected for tidal changes. Dark colors indicate deeper water while lighter colors represent shallow water.

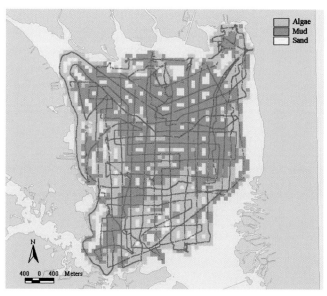

Figure 14. After import into ArcView, the RoxAnn point data was extrapolated using the nearest neighbor algorithm within ArcView Spatial Analyst. This extrapolation was accomplished using a majority classification with a 100-meter cell size. This approach is being tested to determine whether this produces the most accurate results across the entire bay.

Erroneous bathymetry values are detected by sorting by depth in ArcView GIS. Based on NOAA bathymetry for Rehoboth Bay, any point with a depth greater than 4 meters was deleted. Points between 2 and 4 meters were displayed on a NOAA nautical chart to determine their relationship to shallow water. Values extremely different from the neighboring values (e.g., points in excess of 4 meters deep) were identified as unreliable and deleted. Duplicating points were identified and flagged using an Avenue script that detected repeated data values. Each flagged point was closely examined to determine if it was an erroneously repeating value. Points determined to be falsely repeating values were deleted. In addition, any points outside of the study area (due to glitches in the DGPS) were deleted.

We used statistics to better understand the point data. (It is important to note that the data is spatially correlated.) Initially, this study looked at relationships between the E1 and E2 values using JMP Statistical Software (SAS Institute, Inc.). Using a Tukey–Kramer algorithm, all comparison classes (sample sizes and variances not being equal) that were not significantly different based on their E1 and E2 values were detected. Only two classes were completely indistinguishable: undefined sand and sand. These classes were most likely never intended to be separated, and escaped detection until postprocessing due to the similarities in color. The two classes were joined and labeled sand.

Following the removal of repetitive (erroneous) data points and a refinement of the field calibration, a comparison was made between natural breaks and statistically derived clusters. The data was clustered around four means using a k-means cluster. Four classes were selected because three classes for the benthic cover present in Rehoboth Bay were confidently identified: algae, soft mud, and harder mud and sand. An additional class was added for any unknown bottom types that may have been surveyed.

The 4-means cluster analysis clustered the algae based on E1 (roughness) values higher than the bare classes. The other three clusters were very smooth classes with varying levels of hardness. In general, the four acoustic signature clusters appeared to capture the major observations derived from the field calibration. However, the field calibration may have included more relevant information than is possible to glean from statistical analysis. For example, macroalgae, *Ulva* in particular, were of primary concern in this study. The cluster analysis did not appear to detect differences between sparse algae, smaller algae plants, or flattened sheets of algae. However, these algae were observed during ground-truthing and, as such, were included in the field calibration. For the final analysis, the field-derived classification was supplemented with information gleaned from the cluster analysis.

Prior to data collection, we were expecting to be able to easily differentiate the two algae species based on their differing physiology. During the field calibration, the acoustic signature for the different classes of algae were observed to be wide ranging: from very rough to relatively smooth, and from very hard to relatively soft. Often, apparently monospecific mats of *Ulva* produced variable hardness returns. Substrate type or gas may have influenced the algae data returns. Algae may grow on very soft smooth sediments or hard rough sediments. Due to the small footprint (one-tenth of depth), the sensor may have picked up responses from the substrate in the gaps and patches between the plants. In addition, bubbles of air released during photosynthesis typically produce hard acoustic returns in an area that is expected to be soft and rough. These factors made it extremely difficult to confidently distinguish the two species. Initial calibrations indicated that, with detailed knowledge of the sediment types, it might be possible to narrow the calibration to detect subtle differences. For the final classification, however, these classes were grouped into a general mixed algae class.

MAPPING

Developing a comprehensive map from point data (figure 15) can be a subjective process and may produce misleading results. Adding to the difficulty, RoxAnn sees a footprint of one-tenth of the depth: in Rehoboth Bay, RoxAnn was classifying an area on average no larger than 0.15 meter. This sample size produces a potentially very heterogeneous data set. In areas of known algae, small gaps and patches are common between plants. RoxAnn may have detected these small patches and underestimated the amount of algae present. In addition, RoxAnn may have detected debris, individual rocks, fish, or horseshoe crabs. All of these factors can influence the data values. In order to address this issue, we will use the 2000 aerial photography-derived polygons to help characterize the beds of algae in the shallow water. These characterizations will be used to help delineate algal mat boundaries and develop a standard methodology.

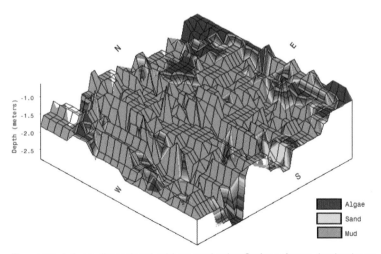

Figure 15. A three-dimensional grid created using Surfer software, having been extrapolated from the RoxAnn point data for a section of Rehoboth. The eastern edge is closest to the shoreline, where the water is shallower. The western edge of the image approaches the center of the bay where the water is deeper. Eventually, this bathymetry data will be integrated into hydrological models to understand nutrient and pollutant movement.

CONCLUSIONS

DCP and all interested parties now have access to high-quality, high-resolution electronic color images as a base layer for later studies and other Rehoboth area work. This project also provided us with significant image-processing experience that will aid DNREC and other organizations in the future. Given additional time and support, we may refine the IMAGINE classification work well enough to further streamline the entire process.

Ideally, this project will serve as a prototype for later resource management activities, as well as a base for later macroalgae biomass and location studies. With further refinement of the image-processing work and integration of other advancing technologies, such as RoxAnn, this project may help integrate new and creative methods into the resource management field.

ACKNOWLEDGMENTS

The following organizations provided technical and/or financial support for this project: The Center for the Inland Bays, the NOAA Coastal Services Center, the NOAA Office of Coastal Resource Management, and Delaware's Department of Natural Resources and Environmental Control.

REFERENCE

Wilkinson, G. G. 1996. A review of current issues in the integration of GIS and remote sensing data. *Int. J. Geographical Information. Systems* 10 (1):85–101.

ABOUT THE AUTHORS

Kimberly Cole received her bachelor of science degree in biology from Ursinus College and continued with graduate work in marine and estuarine ecology at Rutgers University and the University of Maryland. She is presently working at the Delaware Department of Natural Resources and Environmental Control as an environmental scientist for the Delaware Coastal Programs. Kimberly is the coordinator of several coastal resource projects, including Delaware's Coastal Non-point Pollution Control Program, and participates in coastal research of the water quality of the inland bays, as well as the shorebird and horseshoe crab populations of Delaware Bay.

David Carter is an environmental scientist for Delaware Coastal Programs, Delaware Department of Natural Resources and Environmental Control. He is a coordinator for Delaware's Non-point Pollution Control Program and the Delaware Shorebird Migration Monitoring Project.

Chuck Schonder received his bachelor of arts in environmental studies and geography from the University of California, Santa Barbara, in 1994. His undergraduate education included a year at the University of Queensland, Australia, where he studied biogeography and GIS and did coral reef research. In 1998, Chuck received a master of science degree in marine resource management from Oregon State University. He based his master's thesis on a project he completed on the Pacific island of Saipan, which was developing a GIS for improved local coastal management. Upon graduation from Oregon State University, Chuck worked as an environmental scientist for the Delaware Department of Natural Resources Coastal Management Program, where he participated in the work described in this chapter. He now works as a private consultant, environmental educator, and outdoor guide in San Diego, California.

Mark Finkbeiner's academic background includes a bachelor of science in geography and remote sensing from Northern Arizona University (1979), and graduate studies in water resource management at the University of Nevada–Las Vegas. His work in the remote sensing field began with Lockheed Martin Corp. at the Environmental Protection Agency's (EPA) Environmental Monitoring Systems Laboratory in Las Vegas, Nevada. Through the mid-1980s he applied aerial photography to historical characterizations of hazardous waste sites, and spill contingency response planning for EPA and private-sector customers. Since the late 1980s primary tasks included digital land cover mapping using Landsat™ and MSS imagery, and development of remote sensing/accuracy assessment methods to support the EPA's EMAP and North American Landscape Characterization project. Also, between 1988 and 1995 he applied aerial photography to jurisdictional wetland delineations for EPA and the U.S. Department of Justice. Since 1995, he has been at the National Oceanic and Atmospheric Administration's Coastal Services Center where he served as a program coordinator for the Coastal Change Analysis Program. Currently he is a project leader for the center's benthic habitat mapping project, which relies on photogrammetry, videography, and single-beam acoustic sensing to map aquatic habitats on a national basis.

Renée Seaman currently works for the Technology, Planning and Management Corporation (TPMC) at the National Oceanic and Atmospheric Administration (NOAA) Coastal Services Center in Charleston, South Carolina, where she assists with technical methods development, and applies optical and acoustic remote sensing technologies to characterize and map benthic habitats in cooperation with various state-level natural resource agencies. Her academic background includes a master of science in environmental studies from the University of Charleston, South Carolina, and a bachelor of science in biology from Washington College, Chestertown, Maryland.

CONTACT THE AUTHORS

Kimberly B. Cole, Environmental Scientist II
kcole@state.de.us

Contact for further information about Delaware Coastal Programs
David B. Carter, Environmental Program Manager II
dcarter@state.de.us

Delaware Coastal Programs
Department of Natural Resources and Environmental Control
89 Kings Highway
Dover, DE 19901
Telephone: (302) 739-3451
Fax: (302) 739-2048

Chuck Schonder
Sea Dog Solutions
963 Diamond Street
San Diego, CA 92109
scuppers@delanet.com

Mark Finkbeiner, Remote Sensing Scientist
Telephone: (843) 740-1264
mark.finkbeiner@noaa.gov

Contact for further information about the NOAA Coastal Services Center's Coastal
Remote Sensing Program
Renée E. Seaman, Image Analyst
Telephone: (843) 740-1309
renee.seaman@noaa.gov

NOAA Coastal Services Center
2234 South Hobson Avenue
Charleston, SC 29405-2413
Fax: (843) 740-1224
www.csc.noaa.gov

Chapter 6

Geopositioning a Remotely Operated Vehicle for Marine Species and Habitat Analysis

PAUL VEISZE

CALIFORNIA DEPARTMENT OF FISH AND GAME

INFORMATION TECHNOLOGY BRANCH

SACRAMENTO, CALIFORNIA

KONSTANTIN KARPOV

CALIFORNIA DEPARTMENT OF FISH AND GAME

MARINE REGION

FORT BRAGG, CALIFORNIA

ABSTRACT The immense diversity of species and cryptic nature of many marine benthic organisms make the training of properly qualified scientific investigators a long and arduous process. A more viable option for obtaining good, usable data may be to train divers (relatively abundant) in the theory and practice of remotely sensed video imagery, which would in turn be used by personnel trained in algal and invertebrate identification (relatively scarce). With this general goal in mind, the California Department of Fish and Game Marine Region has begun research at the Punta Gorda Ecological Reserve (Humboldt County, California) in the development of non-invasive observation techniques using a remotely operated vehicle (ROV). In October 1998, adverse weather and sea conditions prevented operations in the Punta Gorda area, so an alternate site was selected for the ROV sea trials off Laguna Point, a few miles northwest of Fort Bragg, California. A Deep Ocean Engineering Phantom HD2 ROV, equipped with a video camera and two parallel laser beams spaced 10 centimeters apart, was deployed for measurements of macroinvertebrates and finfish. At regular intervals, the ROV was landed for random recording of fixed-area quadrates, recorded both perpendicular to the substrate with the rotating camera facing downward and horizontally in the case of vertical reef faces. Geopositioning of the ROV was estimated by deploying a Trimble GeoExplorer® II GPS receiver in a waterproof float tethered to the umbilicus of the ROV. A total of 1,100 positions were recorded along a 100-meter ROV transect at a depth of 30 meters during a one-hour deployment. ArcView GIS tables of Coordinated Universal Time and position from the float-based GPS (postdifferentially corrected) were linked to similar tables generated from a shipboard recording of the ROV video camera, which also had GPS-based time code. This technique enabled estimation of the ROV position for any of the video frames of marine organisms captured in the recording. The ArcView GIS hotlink function was then used to display the selected video frames upon spatial selection of ROV positions.

INTRODUCTION

The mission of the California Department of Fish and Game (CDFG) is to manage California's diverse fish, wildlife, and plant resources, and the habitats upon which they depend, both for their intrinsic ecological values and for their use and enjoyment by the public. Seeing is believing, so it is natural for there to be an increasing interest in ocean imagery, particularly as it relates to helping solve complex resource policy problems. In California, recent legislation has focused on nearshore ocean resources. The Marine Life Management Act of 1998, for example, mandated a nearshore fishery management plan statewide within eighteen months. Remotely operated vehicles are seen as a possible tool for safer and more cost-effective information gathering than diver- or manned submersible-based techniques. This chapter describes the CDFG's first experiences with an ROV in a sea trial off Laguna Point, near Fort Bragg, California (figure 1).

Figure 1. Index map of the study area (small rectangle at Fort Bragg) off the northern California coast.

BACKGROUND AND OBJECTIVES

In early 1998, CDFG Marine Region biologists had occasion to view the CDFG Information Technology Branch developments in GPS/ aerial videography. The two units decided to work together to adapt this technology for gathering and recording imagery and other data about the marine environment. The CDFG is following the lead of a number of organizations that have successfully operated ROVs in hopes of reconciling the need for good but hard-to-obtain data with the human safety and logistical cost considerations that surround diver-based biological data collection—notably the Woods Hole Institute in Massachusetts and the Monterey Bay Aquarium Institute, Moss Landing, California. The main objectives of our partnership include:

- mapping of ROV image locations and other sensor data;

- maintenance of a continuous mission time base (from GPS time-driven event log);

- accurate and precise written observations and transcribed video narrative compiled in a digital file, including animal identifications, animal sizes, sample area, substrate/habitat interpretations, environmental factors (sea state, weather, etc.).

METHODS   EQUIPMENT

The ROV (figure 2) was operated from the CDFG's Blue Fin, a 65-foot patrol vessel, with a typical crew of captain, engineer, and boarding officer. ROV sensor and tracking system outputs were recorded with off-the-shelf audio/video components. System integration responsibilities were shared between the ROV supplier (Deep Ocean Engineering, San Leandro, California) and CDFG staff. The stand-alone ROV recording system is hereafter referred to as the ROV/RS.

Figure 2. Photograph of the Deep Ocean Engineering Phantom HD2 ROV.

ROV/RS integrated components consisted of multiple audio and video inputs, GPS/differential GPS (DGPS), a GPS-video encoder, two videocassette recorders, a microphone, a computer display-to-video converter, AC/DC power, and a video/audio monitor, all housed in a rugged equipment tower (figure 3). ROV/RS separate components included laptop computers, a hand-held GPS receiver (Trimble GeoExplorer II) with a waterproof floating enclosure, and a video-still capturing device for the laptop computer. The software used in concert with the ROV/RS included Horita GPSLOGB1.EXE, a video time-code logging program, Trimble Pathfinder Office 2.1 (GPS/DGPS), Microsoft PowerPoint® for video titling, various video capture programs, and ArcView GIS for mapping and visualizations.

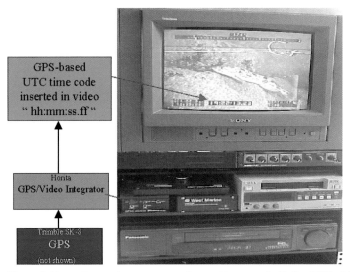

Figure 3. Recording system components used in concert with the ROV: video/audio monitor, with GPS/UTC code stamps imbedded in the video images, a GPS/video integrator, and videocassette recorders.

CONTEXT

Recording of the ROV video was integrated with GPS time and position. This enabled correlation of ROV video events to ROV vehicle tracking and other time-tagged observations. Additional recordable events included voiced observations by pilot or observer or biologist-in-charge, typewritten comments tagged with GPS time and position, and optional, external serial data (such as vehicle depth and heading, sea temperature, salinity, and so forth).

INFORMATION FLOW

A trace of the information flow through the system (figure 4) begins with reception of GPS and DGPS signals through the surface vessel's antennas, and by the internal antenna of the surface buoy-mounted GPS. The shipboard GPS integrates GPS signals and the U.S. Coast Guard beacon DGPS signals to produce real-time, differentially corrected positions and Coordinated Universal Time (the official acronym for which is UTC, also formerly known as Greenwich mean time or GMT). The sea-surface buoy GPS, which approximates the subsea track of the ROV, has its data downloaded after each ROV dive, to be postdifferentially corrected. (DGPS offers submeter positional accuracy.)

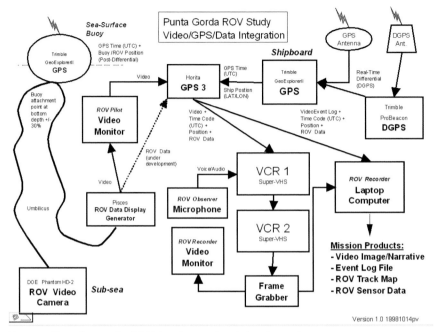

Figure 4. Flow diagram for ROV data integration during the project.

The shipboard GPS sends UTC and position data to the Horita GPS3. This device integrates GPS time and position data with other ROV data (slated for phase II of the study) and generates three separate outputs: (1) a video window of UTC/video frame number; (2) video time code plus position plus ROV data for the right audio track of the VCRs; and (3) a serial data stream sent to the laptop computer for logging mission events, transcribed narrative observations, and so on.

The subsea ROV video camera sends an analog, composite video signal to the Pisces Data Display Generator. This device converts analog voltages from the ROV depth and heading sensors to video character displays (in phase II, Pisces serial data outputs will also be generated). The video stream then passes through the Horita GPS3 where a UTC time display is inserted. The video and all inserted characters are viewed by the ROV pilot and scientist/observer, who have the option to voice remarks and interpretations via the microphone. The video, with all data displays and voice narrative inserted, is then recorded in DVCAM and super-VHS format VCRs (figure 5).

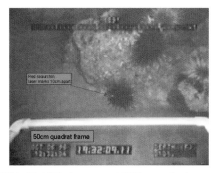

Figure 5. Examples of video imagery from the ROV with UTC stamps and GPS coordinates.

Simultaneously, a third team member, the ROV recorder, operates the video-logging program on the laptop computer. The ROV recorder takes verbal cues from the pilot or observer to mark events, sampling protocol or typing in simple remarks to be associated with a particular video sequence. The log file produced on the laptop is in ASCII text format and tabulates UTC, video frame number, GPS position, and written annotations, as well as mission information such as videocassette ID, location, date, crew, and so forth (figure 6). Correlation of the subsea video with the ROV position is accomplished by matching the UTC time display in a video frame with the UTC time record in the GPS position data file from the surface buoy.

```
┌─────────────────────────────────────────────────────────┐
│ Video logging screen : tag observations with GPS/UTC     │
└─────────────────────────────────────────────────────────┘

LOG                    === Horita PC-LOG(tm) G1.18 ===          Help-F1
------------ PRODUCTION -------------------------- SHOOT ----------------
Title: Punta Gorda ROV Study Dive #2  |Date: 19981006
Number: PGER2                          |Location: Laguna Point (training dive)
Producer: Konstantin Karpov            |Crew: KK;WN;JH;RP;RF;BD;PV(DFG);DM(DOE)
Note: CA Dept Fish and Game Marine Reg |Lat:   0:00.000 N  Lon:   0:00.000 E
--------------- TAPE ---------------- |----------------- LOG ----------------
Reel #: 2               Format: S-VHS  |Date: 19981006  By: pveisze@dfg.ca.gov
Tape #: 2               Length: 120    |Disk #: 1       File: pger2t.log
---------------------------------------------------------------------------
COM1 00:00:00:00   00:00:00  Messages: SETUP...
LINE TC LOCATION   DURATION  COMMENTS                          DROP FRAME
============================ Tabs |T------T--------T--------T--------T----T|=
   8
   9  19:21:13;04  00:38;26  LAT:  39:29.309 N  LON: 123:48.894 W
  10                         obs verm. rf, laser measure
  11  19:21:52;00  00:22;06  LAT:  39:29.309 N  LON: 123:48.894 W
  12                         obs other rockfish ui
  13  19:22:13;17  00:55;28  LAT:  39:29.310 N  LON: 123:48.897 W
  14                         obs lingcod
  15  19:23:10;08  00:29;13  LAT:  39:29.314 N  LON: 123:48.909 W
  16                         appx 15 min tape 1 remaining
  17  19:23:39;21  00:43;26  LAT:  39:29.319 N  LON: 123:48.900 W
  18                         starting ROV return to vessel Blue Fin
Setup-F2  Log-F3  Edit-F4  Get-F5  Save-F6  Review-F7  New-F8  Print-F9  Find-F10
```

Figure 6. Example of video logging screen with observations and UTC time stamps, to be correlated with UTC of positions from float GPS. Coordinates shown in log are position of surface vessel.

Post-dive, the surface buoy GPS data is differentially corrected using shore-based stations. The corrected positions are then input to ArcView GIS running on the laptop computer. The ROV track is then mapped on selectable basemaps. Selected video frames, captured by the frame grabber in real time or from video playback, and other mission data may also be integrated into the GIS for display on screen or plotting to hard copy (figure 7).

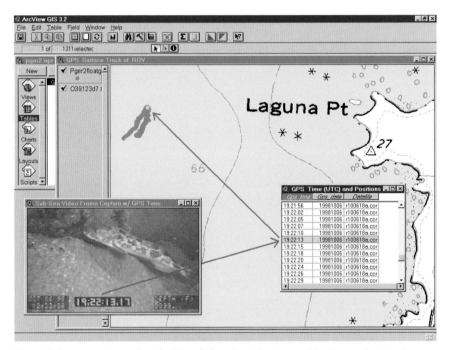

Figure 7. Screen capture of an ArcView GIS project showing a video image that has been hot-linked to the GPS position approximated by the surface buoy attached to the ROV.

ROLE DEFINITION

One of the main considerations in ROV recording operations was the need to dedicate a team member exclusively to the recorder function. Over two or three training cruises, the recorder responsibility evolved into two categories: (1) normal operations, during which the recorder would operate the ROV/RS in response to requests in priority order from the vessel captain, biologist-in-charge, vessel engineer, ROV pilot, ROV engineer/tender; and (2) emergency operations during which the recorder would be responsible for documenting the time, location (latitude/longitude), estimated depth, and description of ROV abandonment.

RESULTS A total of 1,100 positions were recorded along a 100-meter ROV transect at a depth of 30 meters during a one-hour deployment (figure 8). ArcView GIS tables of UTC time and position from the float-based GPS (postdifferentially corrected) were linked to similar tables generated from a shipboard recording of the ROV video camera, which also had GPS-based UTC code. This technique enabled estimation of the ROV position for any of the video frames of marine organisms captured from the recording (figure 7).

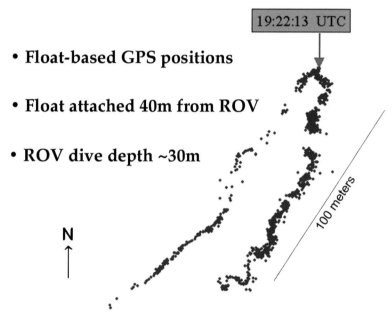

Figure 8. GPS positions mapped over a period of one hour along a 100-meter transect at a depth of 30 meters. Positions were generated using a Trimble GeoExplorer II GPS (post-differentially corrected) encased in a surface buoy attached to the ROV's umbilical.

CONCLUSION ROVs offer an effective means of obtaining non-invasive marine biological information that is safer than diver-based methods and more affordable than manned submersibles. In addition, the amount of information collected by ROVs far exceeds the amount of usable data that can be recorded by divers. For example, in a September 2000 cruise an ROV was successfully used to conduct invertebrate and habitat surveys in the Farallon Islands off San Francisco, an area that is off limits to research divers due to the concentrated population of white sharks.

ROVs also allowed increased bottom time, and certainly an increased depth capability over diver-based methods. During the above survey, an additional calibration cruise was conducted in-shore where both a diver and the ROV were used concurrently to enumerate sea urchins and abalones along a 70-meter line-transect at a depth of 10 meters. Comparable counts were obtained by both methods, but the diver method required forty-five minutes to complete versus five minutes with the ROV. Additional time savings accrue because, unlike divers, the ROV does not require decompression time.

GPS-based time coding of ROV video and audio enables the integration of a wide range of data, as well as a variety of postproduction sampling and information extraction techniques. In the future, the CDFG will use the ArcView Tracking Analyst extension for real-time visualization of ROV positions. This will be especially valuable in study areas such as the Punta Gorda Ecological Reserve and the Gulf of the Farallons National Marine Sanctuary, which have been previously or concurrently mapped with sidescan or multibeam sonar. Seafloor image classification can thus be ground-truthed using the ROV video camera, with the ROV position displayed over the subject seafloor images. The Horita GPS/Video system can also be used to play back recordings and generate position data for use in postprocessing, such as mission simulations.

Additional future projects include the integration of hydro-acoustic, transponder-based systems (such as the ORE TrackpointII), which will enable more accurate positioning of the ROV than the float-based GPS techniques. Using such a method in combination with sonar-based habitat maps, future surveys can be directed to specific habitats. The resulting stratified sampling approach will yield more accurate estimates of species abundance over space and time.

ACKNOWLEDGMENTS   The following individuals and organizations provided technical and financial support for this project: CDFG biologists/ROV pilots Walter Nordhausen and John Hendrix; Blue Fin crew members Robert Puccinelli and Brian Delano; Dirk Rosen of Deep Ocean Engineering (which supplied the ROV); and Robert Pargee of the Horita Company (which served as the GPS/video supplier).

RELATED WEB SITES

Deep Ocean Engineering
www.deepocean.com

Horita Company
www.horita.com

Monterey Bay Aquarium Research Institute (MBARI)
www.mbari.org

Woods Hole Oceanographic Institution (WHOI)
www.whoi.edu

ABOUT THE AUTHORS

Paul Veisze is the California Department of Fish and Game's regional GPS/GIS coordinator. He is also a registered professional forester and licensed commercial/instrument pilot. Paul holds a bachelor of science degree in renewable natural resources from the University of California, Davis, and a master of science degree in forestry from the University of California, Berkeley.

Konstantin Karpov was the ROV project leader for the study and is a California Department of Fish and Game Senior Marine Biologist and Supervisor for northern California. He directs statewide abalone research and management teams, supervises marine field staff from the Oregon border to Sonoma County, and continues as a principal investigator for both the Punta Gorda Ecological Reserve and Northeast Assistance Program research grants to the California Department of Fish and Game. Kon holds a bachelor of science degree in biology from California State University, Los Angeles, and a master of science degree in marine biology from California State University, Fresno.

CONTACT THE AUTHORS

Paul Veisze
California Department of Fish and Game
Information Technology Branch
1807 13th Street, Suite 201
Sacramento, CA 95814
Telephone: (916) 323-1667
Fax: (916) 323-1431
pveisze@dfg.ca.gov

Konstantin Karpov
California Department of Fish and Game
Marine Region
19160 South Harbor Drive
Fort Bragg, CA 95437
Telephone: (707) 964-9078
Fax: (707) 964-0642
kkarpov@dfg2.ca.gov

Chapter 7 The First Three-Dimensional Nautical Chart

CAPTAIN STEPHEN F. FORD

S & J INTERNATIONAL, INC.

WORCESTER, MASSACHUSETTS

ABSTRACT   In the quarter-century between 1976 and 1999, paper nautical charts, like most hard-copy documents, became digital and electronic. They moved as well beyond traditional notions of nautical maps into the realm of GIS. This history is captured from the perspective of one of the creators of the first "electronic chart." The evolutionary vs. revolutionary (Ford 1994) development of electronic chart display and information systems (ECDIS) is reviewed, along with relevant enabling technologies such as the Physical Oceanographic Real-Time Systems (PORTS). The construction of the first three-dimensional nautical charts, made of the Cape Cod Canal region from digitized nautical paper charts and available federal GIS data, is described, with its importance to mariners and coastal zone managers explained. To this end, the use of the ArcView GIS and ArcInfo packages is illustrated as a proof-of-concept. Finally, the need for federal focus on 3-D data is justified with the identification of specific federal agency issues and areas for improvement. The movement of cartography from the portrayal of a scaled, "perceived" or subjective world under the generalization principle, toward an almost 1:1 electronic depiction of an object-oriented, "duplicate electronic world," is underscored. The need for a single, integrated horizontal and vertical datum is stressed as being critical to three-dimensional and four-dimensional cartography in the twenty-first century.

INTRODUCTION   The world's oceans cover 71 percent of the globe and involve a multi-disciplinary group of local, national, and international communities of marine operators, researchers, and administrators, including:

- professional mariners

- port management officials

- fishermen and shellfishermen

- small craft operators, recreational boaters

- coastal zone managers

- marine sanctuary managers

- environmental and natural resource managers

- commercial/recreational fisheries managers

- emergency response personnel

- boundary managers

- oil spill/incident clean-up managers

- environmental modelers

- coastal erosion researchers

- specialists in wetlands and habitat studies

- specialists in coastal/riverine ecosystems

- marine scientists (oceanographers)

- NOAA National Marine Fisheries Service personnel

- U.S. Coast Guard cartographers, navigation managers

- U.S. Geological Survey coastal geologists and watershed scientists

- U.S. Army Corps of Engineers channel maintenance personnel

- Federal Emergency Management Administration (FEMA) personnel (e.g., in hurricane and flood disaster modeling)

- Environmental Protection Agency (EPA) and water quality officials

These groups and individuals interact at the highest legal and operational levels of marine policy, science, and transportation. Unfortunately, their activities and interactions are rarely based upon a common, compatible database, and may sometimes be founded upon incomplete scientific information. Historically, important decisions have been made with the use of two-dimensional maps and systems of limited accuracy and resolution, and with no modeling capability. Oceans are three-dimensional, and in order for these communities to efficiently and effectively carry their individual and collective stewardships into the 21st century, they must have access to high-resolution three-dimensional information. This chapter describes the development of the world's first 3-D nautical chart, which permits spatial modeling over time and is suitable for incorporation into geographic information system (GIS) analysis. Such an extension of GIS could enhance navigational and coastal operations by reducing environmental risks and by reducing the potential for vessel grounding. It will also improve baseline environmental and geophysical data, thereby improving incident planning and response. For example, in the event of an emergency, a properly attributed 3-D nautical chart enables mariners to initiate local management of the incident with appropriate concerns for environmentally sensitive areas.

Traditionally, GISs have been used in the mapping and modeling of two-dimensional, shore-based data for a homogeneous user group that focused on applications relevant to land features (Wright and Goodchild 1997). However, balanced use of ocean resources with environmentally sensitive cohabitation of business, biology, ecology, and the populace is critical as well. Nearly half of the world's population lives within 50 miles of a coastline. This density has created a strong marine transportation business comprising nearly eighty thousand commercial vessels. With the globalization of world trade, waterborne transportation of goods is increasing, with an associated increase in the exposure of coastal zone ecology to maritime incidents. Population pressure on coastal zones continues to rise exponentially. The integration, therefore, of bathymetric and topographic data with nautical charts, digital orthophotos, and other appropriate thematic data is critical to successful management of the coast. GIS has proven an effective tool for coastal zone (e.g., Bartlett 2000) and now marine sanctuary management (e.g., Killpack et al., this volume; Wright, this volume). GIS also works for the captain of a vessel as a way of managing maneuvering and navigational scenarios, or the unfortunate maritime incident. For federal regulatory and policy officials, GIS is an important decision-support tool. At the present time, however, besides knowledge of local geography, the common link between the coastal zone or sanctuary manager, the vessel master, and other regulatory officials is the nautical paper chart.

HISTORICAL BACKGROUND

In 1976, the author conceived and led the development team for NAVSHOALS, the Exxon/Sperry Marine navigation aid module for the Sperry collision avoidance system (CAS). NAVSHOALS provided a geographic overlay with the radar data displayed on a vessel's CAS screen, giving mariners graphical information regarding coastlines, hazards, shoals, traffic lanes, and aids to navigation. NAVSHOALS thus reduced grounding risks and freed mariners to focus on the collision hazards of their maneuvering situation by reducing the time and effort involved in assessing radar target threats.

In 1986, the International Hydrographic Organization (IHO) and the International Maritime Organization (IMO) set out to develop standards and specifications for what was ultimately to become known as the electronic chart display and information systems, or ECDIS (Ward et al. 2000). This was now the mariner's management information system. ECDIS consists of vector electronic charts equivalent in detail to nautical paper charts and integrates all of a vessel's navigation sensors into one system. Similar to NAVSHOALS, ECDIS enables mariners to spend even more time making maneuvering decisions instead of trying to gather the information with which to make the maneuvering decision.

In the early to mid 1990s, Ford et al. (1994) adapted cubit linear measurement theory to three-dimensional (3-D) nautical data collection, processing, and application for multidisciplinary users. Their application incorporated bathymetric, topographic, environmental, and nautical data as an integrated 3-D visual system for mariners, coastal zone managers, environmentalists, administrators, and the like. By 1995 submeter-accuracy electronic docking charts had been developed as an aid to navigation for in-port, reduced-visibility operations. ECDIS had also evolved to higher accuracies of vector chart production, but these were still slowly produced. This situation stimulated the need for a raster nautical chart (RNC), and the electronic chart system (ECS) was born as an interim nautical charting technology for faster production rates.

In 1996, the Physical Oceanographic Real-Time System (PORTS), an application of the Ford et al. (1994) adaptation of cubit theory for 3-D marine data collection and dissemination, was installed and initially operated in Galveston Bay (Ford and Bald 1997). PORTS is a system of five water-level gauges, three current meters, and meteorological stations that provides real-time tide and current data, corroborated with meteorological data every six minutes, for the 50-mile long Galveston estuary. The system also facilitates the input and editing of real-time water level displays in onboard, real-time, 3-D chart display units. Because tidal levels and currents are strongly influenced by local conditions, the advantage of such real-time data to mariners, fishermen, coastal zone managers, and hazardous response teams is immeasurable. A PORTS installation goes a long way toward ensuring the safety of life, property, and natural habitats, both on the water and onshore. Other, smaller PORTS systems have been installed in New York and San Francisco (e.g., see www.co-ops.nos.noaa.gov), with additional PORTS installations under consideration by port authorities across the country.

The recent development of complementary technologies has brought the 3-D nautical chart concept to the edge of technical feasibility, at the cost-effective level of desktop computers. These enabling technologies and advancements include, but are not limited to:

- shallow water multibeam sonar

- raster, vector, and hybrid 2-D electronic charts

- worldwide, real-time satellite communication transfer of data

- advancements in PC computer hardware/software

- the differential Global Positioning System (DGPS)

DGPS now permits submeter-level accuracy in both the horizontal and vertical positioning of data points (e.g., Remondi 1992). Shallow-water multibeam sonar now permits acquisition of dense, near-shore, and harbor bathymetry data. Satellite systems permit near-real-time transfers of data. Modern software permits the combination of raster and vector data. PC hardware is approaching the capability of handling large oceanographic data sets of gigabyte capacity. The 3-D nautical chart, consequently, is now considered a feasible proof-of-concept whose time has come.

STUDY AREA

The Cape Cod Canal was chosen as a study area for its prominence as a principal U.S. east coast water transportation corridor and the abundance of high-quality digital data available for the region. The canal is a deep-draft (32 feet deep and 400 feet wide) sea-level waterway connecting Buzzards Bay with Cape Cod Bay. The waterway measures 15 miles from the Cleveland East Ledge Light to Cape Cod Bay with significantly varying bathymetry and topography. It is routinely used by large vessels of 500 to 650 feet in length. In addition, three bridges with a vertical clearance of 135 feet cross the canal. The canal shortens the distance between points north and south of Cape Cod by 50 to 150 miles and provides an inside passage for avoiding the Nantucket Shoals. Near the canal's northern entrance, vessel traffic routes cross a critical habitat for northern right whales. The mean tidal range is 3.5 feet and the canal has very significant tidal currents of 4.0–4.5 knots. The locale regularly experiences fog and ice floes. In summary, the canal is a critical area for prudent navigation that qualifies it as a site suitable for testing the development of a 3-D nautical chart. Indeed, there are many benefits for the mariner to be derived from a peeling back of the water layer which, perforce, obstructs the view of underwater bathymetry. Rather than dealing with the 2-D nautical charts and mentally constructing a 3-D visual model (figure 1a), with a 3-D nautical chart the mariner can more easily "see" underwater hazards and obstructions (figure 1b), and thus maneuver the vessel in a fashion similar to the way an automobile is driven.

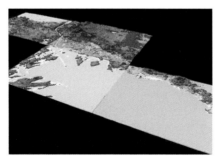 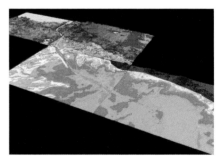

Figure 1. Left panel (1a) shows a USGS DEM of the Cape Cod Canal, Buzzards Bay entrance, showing topography and water. Right panel (1b) shows the same Buzzards Bay topography with additional bathymetry from NOAA.

DATA COMPILATION   Bathymetry data was obtained from the National Ocean Service (NOS) of the National Oceanic and Atmospheric Administration (NOAA), topographic data sets from the U.S. Geological Survey (USGS), and U.S. Army Corps of Engineers (ACoE) air photos from the Massachusetts Geographic Information System (MassGIS). Each data set was in a different original format and required different initialization procedures in order to input them to ArcView GIS. The USGS data had to be untarred and passed through a Spatial Data Transfer Standard (SDTS) filter prior to importing to ArcView GIS. The NOAA data sets were derived from the surveys used to compile the bathymetry for NOAA nautical charts 12320 and 12326. This data was directly imported to ArcView GIS as a digital elevation model (DEM). Data sets from both the NOAA and the USGS were accompanied by metadata (data about data) to facilitate the uploading of the respective data sets. Due to their size and complexity, the federal data sets required two and a half times the amount of production time as was required by digitizing paper maps.

Figure 2 shows the southwest and northeast entrances to the Cape Cod Canal (Buzzards Bay and Massachusetts Bay, respectively) and illustrates the interfacing responsibilities of the federal agencies involved in bathymetric and topographic data collection and dissemination in this region. Prior to releasing elevation and depth data to the public domain, the federal agencies submit the field data to extensive processing and quality control checks. In this process, a typical bathymetric survey of two to five million soundings will be dramatically reduced to a point file of approximately a million. NOAA's depth data is converted to contours prior to public release. This process smooths the actual data and inserts the equivalent of an "artifact" into real-world data as the real world is not comprised of smoothed "plateaus." Therefore, because of the way the process preserves point values, a 3-D nautical chart partially digitized from paper charts may be more representative of reality than the "contoured" federal data sets.

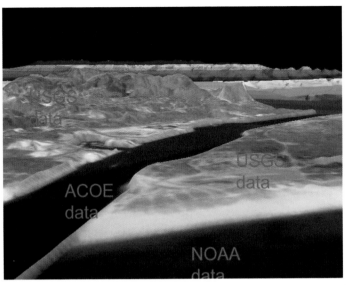

Figure 2. A 3-D airphoto of the northern Cape Cod Canal, Massachusetts Bay entrance, made with source data from NOAA, the USGS, and ACoE.

ACCURACY AND ERROR ISSUES

In terms of geographic coverage, there was significant overlap between the NOAA and USGS data sets, both of which were in the same horizontal datum (NAD27), but with different vertical datum references that had to be reconciled and standardized prior to combination. The NOAA data had greater vertical resolution and accuracy than the USGS data. The NOAA data, however, contained elevations greater than zero; these points had to be removed in order to delineate a shoreline and the nautical chart datum reference line of mean lower low water (MLLW). Bathymetric depths on a nautical chart are referenced to the local tidal datum, which is typically MLLW-averaged over the nineteen-year tidal epoch. Depths below the datum are negative and elevations above are positive. The vertical datum for the depths was NGVD29. The 1:70,000 scale of the NOAA medium-resolution digital vector shoreline was not deemed an acceptable level of detail for this project, due to the 1:40,000 scale of source raster charts. The NOAA metadata cites a locational accuracy for the soundings of 0.0015 meters at the source "smooth sheet" level and 3 meters at the paper chart, user level. The vertical accuracy of the paper chart data is cited as 2 percent of depth or 0.2 meters for depths shallower than 20 meters and within 1 meter for depths greater than 20 meters.

NOAA DATA

The NOAA data was available in both a single 30-meter UTM grid for the entire Buzzards Bay estuary, or in individual, randomly dispersed data sets for portions of the estuary. Because of its size, our hardware and software were not able to handle the single, large grid for Buzzards Bay along with the additional project overlay themes. The UTM grids did not align with the typical geographic parallels and meridians, which define the extent of the USGS 7.5-minute DEMs. Hence, edgematching with other sources will be a future issue.

A vertical root mean square error (RMSE) statistic that includes both random and systematic errors is used to describe the vertical accuracy of a DEM. The vertical accuracy of the constructed NOAA bathymetric DEMs is estimated to be 2 percent of depth (i.e., 1 meter for depths greater than 20 meters and 0.2 meters for depths less than 20 meters). Because of three types of DEM vertical errors (blunder, systematic, and random), the NOAA metadata carries a warning that the data "should not be used for navigation purposes."

Blunder errors are errors of major proportion and are easily identified and removed while editing the data. Systematic errors are "fixed-pattern" in nature and are introduced by data collection procedures and systems (e.g., unsampled elevation shifts, relative spacing of the source soundings, surface misinterpretation due to "soft" reflectivity and fractional sounding resolution). Random errors result from unknown or accidental causes. All of the DEM errors can be eliminated by good quality control processes. Also, the NOAA bathymetric data exceeds National Mapping Standards by a factor of three, making it significantly more accurate than the USGS data, which meets but does not exceed those standards. After consideration of the "warning," the author's analysis of the data found it to be of a quality suitable for the purpose of developing a proof-of-concept 3-D nautical chart, particularly because the 3-D chart also incorporates the accuracy of NOAA's paper nautical charts.

USGS DATA

The USGS data was provided as 30-meter, 7.5-minute DEMs, in UTM Zone 19 with a NAD27 horizontal datum and a NGVD29 vertical datum. The National Mapping Standard of 7-meter RMSE or an RMSE less than the contour interval is what the USGS uses as the standard for its DEMs (USGS 1997). The USGS shoreline reference is mean sea level (MSL) and causes a discontinuity equal to a small percentage of the tidal range with respect to the MLLW datum of NOAA.

The resulting 3-D nautical chart project used MLLW as the tidal datum on a Mercator projection and a WGS1980 (NAD83) spheroid with a vertical NGVD29 datum. The digitized version of NOAA Nautical Chart 13236 was executed to these parameters. The federal data sets had to be projected to these new parameters from their original UTM format.

3-D NAUTICAL CHART PRODUCTION

The following procedures were used to create a 3-D nautical chart for the Cape Cod Canal:

1a Digitize paper NOAA nautical chart 13236 of Cape Cod Canal.

1b Construct a triangulated irregular network (TIN) from digitized coordinates.

1c Georeference and overlay a NOAA digital raster version of chart 13236 on TIN (figures 3, 4, 5).

1d Overlay air photos and other thematic data from the region covering chart 13236.

Steps 1a–d resulted in one 3-D NOAA nautical chart, 13236, in Mercator projection constructed from both the paper and digital raster charts.

Figure 3. NOAA raster nautical chart 13236 of Cape Cod Canal (with navigational aid symbols) overlain on TIN constructed from digitized paper version of chart (vertical exaggeration present).

Figure 4. Enlarged view of portion of scene from figure 3 showing topography, islands, and underwater channels, along with navigational aid symbols (vertical exaggeration present).

Figure 5. Enlargement of central portion of figure 4 at the turn into the main inbound channel, just below the southern end of a dike in Cape Cod Canal (vertical exaggeration present).

2a Merge USGS DEM with NOAA bathymetric grids.

2b Construct a second TIN from merged data sets.

2c Georeference and overlay NOAA digital raster version chart 13230 of Buzzards Bay on second TIN (figure 6).

2d Overlay air photos and other thematic data from the region covering chart 13230.

Steps 2a–d resulted in one 3-D NOAA nautical chart, 13230, in UTM Zone 19 projection for coastal zone managers, and in Mercator projection for mariners.

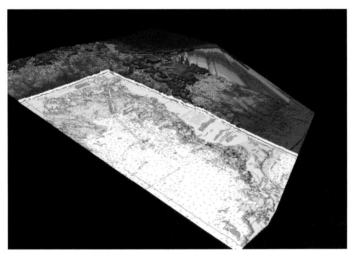

Figure 6. Overlay of NOAA raster nautical chart 13230 of Buzzards Bay on TIN constructed from USGS DEM merged with NOAA bathymetry.

3a Overlay and clip data coincident with chart 13236 (Cape Cod Canal) from second TIN (Buzzards Bay) to create third and final TIN (figure 7).

3b Georeference and overlay NOAA digital raster version chart 13230 (Buzzards Bay) on third TIN (figure 8).

3c Overlay air photos and other thematic data from the region covering chart 13236 (Cape Cod Canal).

Steps 3a–c resulted in one 3-D NOAA nautical chart, 13236, clipped 3-D nautical chart 13230 in Mercator projection.

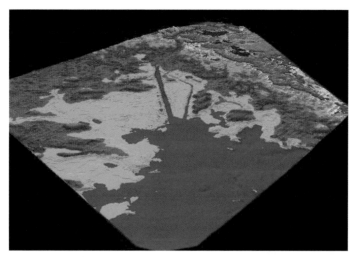

Figure 7. A TIN for region covered by NOAA raster nautical chart 13236, clipped from Buzzards Bay TIN constructed in step 2b.

Figure 8. Overlay of NOAA raster nautical chart 13236 (Cape Cod Canal) on Buzzards Bay TIN constructed in step 2b.

INSERTION OF REAL-WORLD OBJECTS INTO 3-D NAUTICAL CHARTS

It is readily evident that the use of 3-D objects as aids to navigation will reduce the amount of text required on a raster chart image, and will reduce as well the amount of time and effort needed by mariners to identify and interpret navigational aids. This, in turn, reduces the risks of an incident due to faulty navigation and increases the amount of time a mariner can spend "looking out" (the best collision reducer).

The use of real-world 3-D objects to symbolize lighthouses, bridges, and similar navigation subjects was an unattained goal of this project. Due to the limitations of the selected software, only 2-D views of the constructed 3-D objects were exportable to the 3-D nautical chart. However, figures 9 and 10 convey the concept with two-dimensional objects for the southwest and northeast entrances to the Cape Cod Canal. It is anticipated that the next-generation 3-D nautical charts will include three-dimensional representations of these navigational reference points.

Figure 9. Enlargement of a portion of a 3-D nautical chart (on the northern Cape Cod Canal, near Massachusetts Bay) with images of navigational aids and a ship inserted into the scene.

Figure 10. Enlargement of a portion of a 3-D nautical chart (on the southern Cape Cod Canal, near Buzzards Bay) with images of navigational aids inserted into the scene.

In the current 3-D nautical chart application, the aid-to-navigation attributes can be queried singly or as a group, and the inquiry will generate an on-screen attribute table with name, characteristics, and location of the queried object or line in a 2-D display window. In 3DSMARTCHART (see below), visual characteristics will be readily apparent and it will not be necessary to list them in a table.

3DSMARTCHART

The ultimate goal of the 3-D nautical chart project is the production of an integrated, real-time navigation system, which is called the 3DSMARTCHART. It will accurately reveal the seafloor features to the mariner in 3-D, enabling the mariner to fly the vessel among the hazards, obstructions, and perils of the depths, with the added benefit of real-time sensors for water level, tidal current, and meteorological data (figures 11 and 12). Coastal zone managers may use the system to obtain real-time imagery for disaster mitigation and response.

Figure 11. Enlargement of a portion of a 3-D nautical chart (on the southern Cape Cod Canal, near Buzzards Bay) overlain with planned parameters for a 3DSMARTCHART extension to the current 3-D nautical chart application.

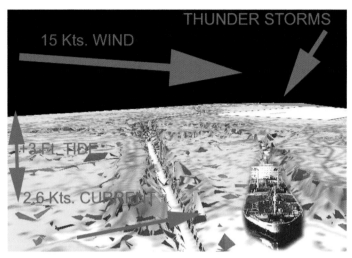

Figure 12. Enlargement of a portion of a 3-D nautical chart (on the southern Cape Cod Canal, near Buzzards Bay) overlain with examples of physical oceanographic and meteorological information that would be available in 3DSMARTCHART.

RELEVANT COASTAL ZONE APPLICATIONS

Three-dimensional charting is not limited to the nautical. The tools can be applied to generate 3-D topographic map overlays and air-photo overlays for a host of scientific and managerial applications. Figures 13, 14, and 15 illustrate 3-D applications for the Scraggy Neck locale of Buzzards Bay showing a wealth of visual geospatial information on coastal dynamics and population pressures. Figure 13 is a USGS topographic map overlain on 3-D topography. Figure 14 shows a color USGS orthophotoquad and two coastal zone themes overlain on 3-D topography. The pinkish polygons identify ecologically sensitive areas. The yellowish polygons identify barrier beach areas. In Massachusetts, the coastal zone extends from the 3-mile limit of the state territorial sea to 100 feet beyond the first major land transportation route encountered (i.e., road, railroad track, and so on). Figure 15 is an enlargement of figure 14, illustrating the richness and detail of the dynamic information found in a 3-D portrayal of spatial data.

Figure 13. Overlay of a USGS topographic map on 3-D topography, Scraggy Neck portion of Buzzards Bay.

Figure 14. Overlay of a USGS digital orthophotoquad on Buzzards Bay TIN constructed in step 2b of 3-D nautical chart production. Polygons were digitized from the USGS topographic map in figure 13: pink signifies ecologically sensitive areas and yellow identifies barrier beaches, Scraggy Neck portion of Buzzards Bay.

Figure 15. Enlargement of figure 14 to show detail available in USGS digital orthophotoquad.

Coastal zone management concerns such as erosion/accretion, habitat locales, nonpoint pollution, marshland, biologic distributions, and so forth, and their respective interrelationships can be dealt with via data themes for variables above and below the water's surface. Other benefits of 3-D nautical charting for coastal zone and marine sanctuary management include:

- plotting and monitoring coastal ecosystems and their physical parameters

- plotting and monitoring coastal population effects and regulatory impacts on coastal communities

- plotting and monitoring biological and fish habitats of commercial/recreational fisheries

- plotting, monitoring, and assessing sea grass habitat and health

- plotting, monitoring, and forecasting coastal erosion and flood levels

- assessment of artificial reef benefits and growth rates

- measuring, assessing, and modeling port development and dredging activities

- plotting, monitoring, and assessing documented U.S. disposal sites

- oil spill response strategy and modeling benefits

- modeling of coastal watersheds and coastal currents and tidal estuaries

- modeling of dynamics and interrelationships of coastal zone parameters

- modeling and managing aids-to-navigation applications, maintenance, and location

- displaying legal and statutory boundaries for operations and litigation

CONCLUSIONS

Geography is global. There is no physical barrier between the land and sea. Nor does a barrier exist at the air–sea interface. Land–sea and air–sea are just benchmarks in the horizontal and vertical continua of nature that begins in the depths of the Marianas Trench and ends at the heights of Mount Everest. The world of two-dimensional charting must grow to incorporate these global realities of three and four dimensions. A modern marine GIS with a core 3-D nautical chart could become the first step in the creation of a managerial tool that can more effectively protect and preserve our precious coastal environs. Lessons learned from the implementation of established maritime technologies such as GPS, CAS, very-high-frequency (VHF) radio, and automatic radar plotting aid (ARPA), will help to improve 3-D nautical charts as well.

Finally, the following recommendations are offered for the attention of federal agencies and software vendors.

FEDERAL AGENCIES (NOAA AND USGS)

1 For 3-D applications, NOAA must shift to a charting policy of a single reference plane (MLLW tidal datum) for the measurement of both charted depths and charted heights. In the interest of vessel safety, MLLW is the most appropriate reference datum. The practice of measuring depths from MLLW and heights from mean high water (MHW) is now an item for the historical past. The elevation zone between MLLW and MHW can no longer be ignored.

2 To facilitate merging elevation/depth data sets for 3-D applications under the existing federal organization, NOAA must defer to the USGS for elevations above the MLLW (zero) datum and maintain its data sets "clean" of any elevations greater than zero. An option is to use the national shoreline as the interface between the areas of agency coverage responsibilities.

3 In support of navigation safety, the USGS must raise its topographic mapping standards for coastal zones to the equal of NOAA's bathymetric standards. The existence of two different mapping standards underlying data in ecosensitive coastal environments is problematic.

4 To maintain consistency of the elevation spectrum, USGS must shift its vertical datum to NOAA's MLLW datum and eliminate the "USGS–NOAA gap," as well as the gap between the MLLW and MSL datums.

5 In the short term, federal policy directives should be executed to ensure that NOAA and USGS are mapping to the same standards and playing by the same rules with respect to data and metadata.

6 NOAA should avoid small "patch-like" bathymetric data sets in estuaries. These small data sets should be integrated into the larger, local sets. The use of version number or edition number can then be utilized to validate the currency of the data set, and small data sets will neither be overlooked nor create software "snags" when being processed.

ACKNOWLEDGMENT

The author gratefully acknowledges the support and contribution of MassGIS (20 Somer Street, 3d floor, Boston, Massachusetts 02108, www.state.ma.us/mgis/massgis.htm), which supplied orthophoto-quads and thematic data for the project.

REFERENCES

Ford, S. F. 1994. ECDIS: Revolutionary or evolutionary? *Proceedings of the National Technical Meeting of the Navigation Institute.*

Ford, S. F., and R. J. Bald. 1997. Houston/Galveston safe passage into the 21st Century. *Proceedings of the Seventh International Offshore and Polar Engineering (ISOPE) Conference,* 297–303.

Ford, S. F., et al. 1994. *Final Report of the Committee on Modernizing Navigation Services.* NOAA Strategic Plan Advisory Committee internal report.

Bartlett, D. J. 2000. Working on the frontiers of science: Applying GIS to the coastal zone. In *Marine and Coastal Geographical Information Systems,* Wright, D. J,. and D. J. Bartlett, eds. London: Taylor & Francis.

Killpack, D., A. Hulin, C. Fowler, and M. Treml. 2001. Protected areas GIS: Bringing GIS to the desktops of then national estuarine research reserves and marine sanctuaries. In *Undersea with GIS,* D. J. Wright, ed. Redlands, California: ESRI Press.

Remondi, B. W. 1992. Real-time centimeter-accuracy GPS for marine applications. *Proceedings of the Fifth Biennial National Ocean Service IHO Conference.*

U.S. Geological Survey. 1997. *Standards for Digital Elevation Models.* U.S. Geological Survey, National Mapping Division document.

Ward, R., C. Roberts, and R. Furness. 2000. Electronic chart display and information systems (ECDIS): State-of-the-art in nautical charting. In *Marine and Coastal Geographical Information Systems,* Wright, D. J., and D. J. Bartlett, eds. London: Taylor & Francis.

Wright, D. J. 2001. Seafloor mapping and GIS coordination at America's remotest national marine sanctuary (American Samoa), In *Undersea with GIS,* Wright, D. J., ed. Redlands, California: ESRI Press.

Wright, D. J., and M. F. Goodchild. 1997. Data from the deep: Implications for the GIS community. *International Journal of Geographical Information Science* 11(6):523–28.

FURTHER READING

Konda, Ford, Landphar, Larsen, and Basillato. 1997. *Proof of Graphic Concepts for Ship Piloting and Docking in Restricted Visibility Conditions.* SWUTC-98/467407-1.

Klepsvik, J. O., and M. L. Bjarnar. 1996. Laser-radar technology for underwater inspection, mapping. *Sea Technology* 37(1):49–52.

Lindholm, M., M. E. Wester, and J. Lehtosald. 1996. Tools for support decision making in navigation with integrated bridge systems. *Institute of Navigation Proceedings of the 52nd Annual Meeting,* 335–41.

Lockwood, M., and R. Li. 1995. Marine geographic information systems: What sets them apart? *Marine Geodesy* 18(3):157–59.

Maloney, F. W.; J. P. Clarke, RADM; R. J. Willis, Commodore; D. Law, Captain; L. P. van der Poel, Captain; and P. F. K. Usher, Commander. 1997. The case for raster nautical charts. *Sea Technology* 38(6):10–15.

Minkel, D. H., LCDR, and G. Whitney. 1992. Implementation of differential GPS at the Coast and Geodetic Survey. *Fifth Biennial National Ocean Service IHO Conference,* 31–36.

Smith, W. H. F, and D. T. Sandwell. 1997. Global seafloor topography from satellite altimetry and ship depth soundings. *Science* 277:1,956–62.

Stembel, O. E., and M. J. Yellin. 1999. Raster nautical charts: Effect of scan resolution and screen resolution on the quality and scale of the chart. Electronic journal: chartmaker.ncd.noaa.gov/ocs/rnc/newpaper.htm.

Whitman, D. C. 1996. Laser airborne bathymetry: Lifting the littoral. *Sea Technology* 37(8):95–98.

ABOUT THE AUTHOR

Captain Stephen F. Ford is a 1970 graduate of the U.S. Merchant Marine Academy with a 1978 MBA from the University of Houston. He is a Master Mariner of fifteen years' seagoing experience. In addition, he is past vice president of a U.S. flag shipping company of nine vessels, and from 1988 to 1996 was the head of the Marine Transportation Department of Texas A&M University at Galveston. He served three years as the PORTS Manager of the Houston/Galveston system, eight years on the U.S. Coast Guard Navigation Safety Advisory Committee (NAVSAC), twelve years as member of the Houston/Galveston Navigation Safety Advisory Council (HOGANSAC), and four years on the Houston Pilot Review Board.

In the mid-1970s Captain Ford developed the first electronic chart that was the prototype for today's modern electronic nautical charts (ENCs) and the electronic chart display and information system (ECDIS). In the 1990s he proposed the development of "smart charts" for the second-generation ECDIS units. This concept includes real-time sensor inputs to vector ENCs in order to give mariners a management information system onboard their vessels. Selected as one of ten key maritime industry personnel, he participated in a 1993 NOAA initiative entitled "Modernizing Navigation Services" to guide senior government leaders in their preparation for the 21st century.

In 1998, Captain Ford obtained a second master's degree in geographic information systems from Clark University. In addition to his professional activities, he has over a dozen publications on a variety of maritime technology topics.

CONTACT THE AUTHOR

Captain Stephen F. Ford
S & J International, Inc.
111 Woodland Street, Suite 12
Worcester, MA 01610
Telephone: (508) 799-6933
captain@splusnet.com

Chapter 8

HIS: A Hydrographic Information System for the Swedish and Finnish Maritime Administrations

KATARINA JOHNSSON
T-KARTOR SWEDEN AB
KRISTIANSTAD, SWEDEN

VELI-MATTI KIVIRANTA
NOVO MERIDIAN LTD.
ESPOO, FINLAND

ABSTRACT

A hydrographic information system (HIS) was created to manage the hydrographic data of the Swedish and the Finnish Maritime Administrations. Descriptions of hydrographic information management include metadata management, feature management, multiple scales management, and history management. A primary objective was the creation of an efficient electronic nautical chart production line, with cell management capabilities such as cell definition and new-cell/revised-cell production. Workflow control was implemented for all modifications to the database. The database was defined in geographic coordinates, and import/export functionality included datum conversion and projection transformation. Advanced editing, linking, and quality control ensured data integrity.

INTRODUCTION

The hydrographic information system developed for the Swedish and Finnish Maritime Administrations is a single system for capturing, managing, and controlling the quality of hydrographic data. Information management, data edit, quality control, and electronic nautical chart (ENC) production are some of the core capabilities of the HIS. The quality of data used by national hydrographic organizations in Scandinavia is a main concern. Starting with the capture of data or import of same from existing systems, HIS manages the flow, capturing information that is essential to verifying the acceptability of a data set. HIS meets the high requirements put on data structures (topology, attributes, and such) by national standards set for nautical chart production. And because the increasing use of GPS for navigation leads to high demands for positional accuracy in charts, HIS also has the tools necessary to move and position objects precisely in unprojected as well as projected data sets. The system design of the HIS described in this chapter was a joint effort between the Swedish and the Finnish Maritime Administrations, Novo Meridian Ltd. (Finland), and T-Kartor Sweden AB, with support on specific technical issues from ESRI.

EVOLUTION OF THE HIS

The introduction of a new hydrographic data processing system is an evolutionary process that requires both organizational and methodological changes, changes to existing systems, and the addition of totally new systems. The implementation strategy of HIS is therefore gradual. The three main phases are defined here.

HIS PHASE 1 (THE SYSTEM)

Evolution of the HIS begins here. The main parts of phase 1 are information management and process control (figure 1). There are interfaces to existing source data and production management systems. Selected parts of source data management and product management are also included in the system.

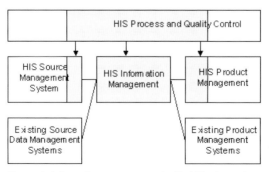

Figure 1. Information management with HIS phase 1.

Integrity of data is a central concern here. All features are stored and maintained in one place, and all features have a unique ID. Data structures and processing rules of the features are the same. Dependencies and links between the features are controlled, and both internal quality and the quality between the features are assured. The number of interfaces between existing systems is minimized as well. HIS therefore facilitates information management for the user because both the user interface and the way of working are uniform. Software and hardware tools are standardized, making maintenance of the system more efficient.

HIS PHASES 2 AND 3 (THE FUTURE)

Phase 2 brings us to full functionality by implementing a cartographic production system that retrieves data from the HIS databases. In phase 3 (figure 2) the system is in full-scale operation. All source data management systems are integrated to the system. Preprocessing of various types of measurements are handled in the HIS, and many other systems with navigational databases can be moved to the same architecture.

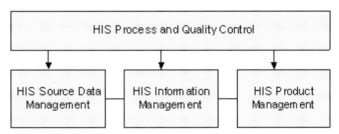

Figure 2. HIS phase 3.

SYSTEM OVERVIEW

CLIENT–SERVER ARCHITECTURE

HIS has a client–server architecture, where Windows NT clients are connected to a UNIX server (figure 3, table 1). The client–server setup allows sharing of data sets, workflow control, supervisor distribution of tasks to technicians, centralized data archiving, and optimized client data processing. ArcSDE and Oracle® software resident on the server store the master databases from which clients extract hydrographic data for updates or ENC production, or to which the clients place newly input and validated data sets (table 1). SDE allows for efficient management, search, and retrieval of geographic data inside a relational database management system (RDBMS). The RDBMS ensures data integrity and provides the security required by the maritime administrations. Staffware, a commercial software for workflow management, handles system workflow control. ArcView GIS on the client side provides tools for hydrographic data editing, viewing, querying, importing, and output. HIS for the Finnish and Swedish Maritime Administrations is designed for up to eighty simultaneous client users (fifty viewers and thirty operators).

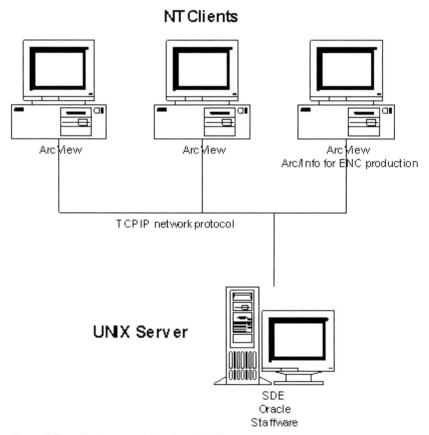

Figure 3. The client–server architecture of HIS.

TABLE 1. HIS HARDWARE AND SOFTWARE

	Hardware	Software	Comments
Server	Digital™ Alpha™ UNIX worksta- tion	ArcSDE 3.0.1 for Oracle	The system can run on any RDBMS supported by SDE.
Workflow		Staffware 5.2	
Client	Windows NT workstation (minimum RAM 64 MB)	ArcView 3.1	One license for each client. ArcView GIS extensions DBAccess and Dialog Designer are required to run the HIS.
		ArcInfo 7.2.1	One license for each client running ENC production.

FUNCTIONALITY

HIS contains thirteen subsystems (figure 4). Data can be input via direct connections to other hydrographic systems, imported through custom-designed importers, digitized, or manually entered. The HIS registers the new data and tags it so that data history is instantly recallable. HIS data management includes metadata management, multiple scales management, and history management. The goal is to keep a high level of data traceability through all aspects of data processing.

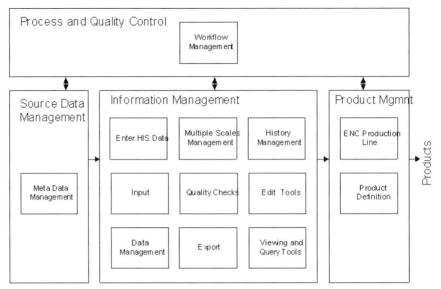

Figure 4. HIS subsystems.

CLIENT ARCHITECTURE

The client application was designed for flexibility and modularity, to allow for easy customization to the needs of other similar organizations. The application was built as a set of more than fifty extensions to the standard suite of ArcView GIS functions. The client side of HIS was built on a flexible framework for multiple language management. Support for three languages was implemented: English, Swedish, and Finnish. This can easily be extended to include other languages. The graphical user interface (GUI) of the HIS client is based mainly on the ArcView GIS user interface, which provides a Windows look-and-feel to menus, buttons, tools, and dialogs, with context-sensitive help functionality. The GUI adapts to the current tasks of the operator, and only those functions that are applicable at certain stages of processing are actually available.

DATABASE   DATABASE DESIGN

The core of the HIS is the database, which consists of the main database and several scale databases (figure 5). Multiple-scale databases are required for the system to maintain accurate base data, with the geographic database, as well as generalized data, suitable for presentation on specific scale ranges of cartographic databases. The number of databases at various map scales is not fixed, and the system can support any reasonable number of them (typically three to ten).

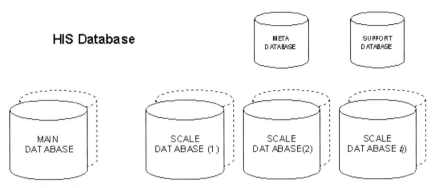

Figure 5. The HIS database.

The main database (the geographic database) contains data at its most accurate level. Normally, new data of good quality is first entered into the main database, then transferred to one or many scale databases. The scale databases (the cartographic databases) contain data generalized for cartographic output at a certain scale range. The object types, their attributes, methods, and logical and topological relationships, as defined by the Swedish/Finnish Offices data model, are implemented in these databases. On the client side, the database structure is described in an HIS data dictionary, to allow for application code that is largely data independent. Metadata is stored in a metadatabase, and supporting information, such as logs, are stored in the support database.

The HIS database is implemented as an Oracle and SDE instance, while the internal databases (main, scale(n), support, meta) are implemented as Oracle user schemas. The implementation of the HIS database and the HIS data dictionary was guided by the use of Geographic Business Systems' modeling software GeoCASE.

FEATURE OBJECTS

The following feature object types are supported:

- simple objects: spatial without attribute-level relations to other objects;

- complex objects: spatial or nonspatial with attribute-level relations (links) to other objects; and

- composite objects: objects with parent–child relationships where the child objects cannot exist without their parent objects.

The feature object classes fall into four categories (table 2).

TABLE 2. CATEGORIES OF FEATURE OBJECT CLASSES.

Navigational features	Skin of the earth	Soundings	Other objects
Navigational aids and their components	Depth areas	Significant soundings	Buildings
Navigation lines	Generic (land) areas	Sounding clusters	Bridges
Fairways and fairway areas	Depth contours	Surveys	Transmission lines
Limits	Coastlines		All other simple object classes
Rocks	Closing lines		

HISTORY MANAGEMENT

HIS supports historical viewing and querying of data via history tables. History tables are duplicates of the tables in the main database and the scale databases, with additional fields for start date and end date of the validity of the feature. Whenever an object in the main or scale databases is edited, the previous versions of the object are copied to its history table. This allows the application to reconstruct the situation of the database at a certain point in time by consulting both the main and historical tables. It allows the user to track the changes made to a specified feature.

LONG TRANSACTIONS

Due to the length of time that may be required to perform complex edits of spatial features in the SDE database, long transactions are supported in HIS. This is to prevent users from copying out features for edit that have already been copied out for editing by another user but have not yet been returned to the database. Long transactions are handled on an individual feature basis, locking the features that are being edited, as well as all related features. Avoiding conflicts resulting from two operators working in the same area but with different data is a task for the system supervisor. This is facilitated via the concept of rough polygons, which outlines the geographical extent of any one ongoing job. A long transaction starts with a checkout of data from the database and ends with a "commit" of data to the database. A commit itself is a short transaction. It is also an atomic operation, which means that either all or no data goes into the database, thereby avoiding inconsistencies in the data.

QUERYING AND VIEWING HIS DATA

HIS offers versatile functionality to view and query features, feature metadata, and history or processing history of the features. The system supports multiple view management with zoom-in views for full display of details, as well as full cartographic display of data with point, line, and area symbols. The system provides a set of basic symbol libraries, and additional libraries may be easily built. The point symbols are in TrueType® fonts. HIS supports intelligent theme management based on object classes and data types (figure 6). The interface to the user is expressed in terms of feature objects well known to the operator. Object query is handled through flexible dialogs, which handle both object attributes and object relations. Other key features of the HIS view and query management include:

- annotation text settings;

- complex symbol visualization (e.g., for light sectors and depth values);

- tools to measure length and size of objects;

- tools to view the attributes and relations of objects;

- a full set of selection tools; and

- geodetic tools to compute (e.g., errors due to projection, and true geodetic distances).

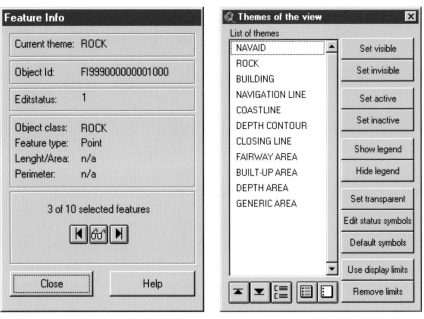

Figure 6. Dialogs for feature and theme management.

DATUM AND PROJECT MANAGEMENT

The HIS database is stored in geographical coordinates. All import/export functionality includes datum and projection conversions as needed. The datum and projection properties of incoming data are stored in the metadatabase and can be retrieved at will. Data in the view can be projected on the fly to any projection known by the system. A system administrator can enter projection parameters for any projection.

EDIT AND QUALITY CHECKS

Object classes are edited one by one. The appropriate line, point, or polygon tools are made available depending on data type (figure 7). For some complex object classes, additional special-purpose geometry tools have been implemented. Most importantly, there are linking tools that allow the operator to build up critical relations between objects (figure 8), for example between navigation lines and their leading marks (navigational aids).

Figure 7. Dialogs for geometrical editing and attribute transfer.

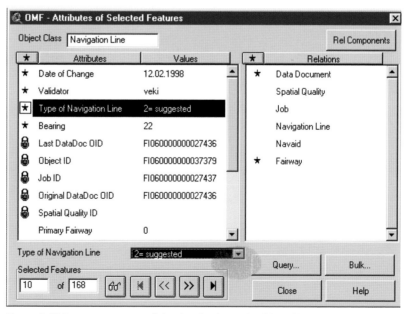

Figure 8. Object management dialog for viewing and editing object attributes and relations.

HIS supports input of features by coordinates, since exact position is crucial for many hydrographic measurements and features. Through the concept of non-draggable objects, graphical drag-and-drop is prohibited for extra-sensitive object classes. HIS has full support for datum and projection conversion during coordinate input. Various methods for coordinate input are supported, such as absolute coordinates, relative coordinates to existing objects, and relative coordinates by distance and bearing to and from existing objects. For "skin-of-the-earth" object classes, holes are not allowed and objects cannot cross or overlap each other. The edit tools can work under skin-of-the-earth constraints, and there are special-purpose tools to fill gaps and check that overlap does not occur. Quality checks have been developed to verify internal feature quality and the quality between the features. The checks can be run against the current workspace or against the database to ensure that attributes, geometry, and topology are correct inside a feature and between features. These quality checks are partly intended as prechecks for the database checks in "commit," but they also handle complex geometric and topology constraints that cannot be implemented on an Oracle level.

WORKFLOW MANAGEMENT

HIS workflow management, or process control, is the computer-assisted management of hydrographic data processing via ArcView GIS software. The order of execution is controlled by Staffware's computerized representation of business processes.

All tasks that involve modifications to the main and scale databases, as well as all ENC production, are under workflow control. Jobtypes—predefined sequences of steps (processes)—have been identified for these tasks. Jobs are instances of jobtypes, and ongoing jobs can be monitored and managed through the Staffware work queue. Work queue management is provided for individual users, user groups, and supervisors.

On the client side, workflow steps are implemented as process extensions that are loaded and unloaded to provide the functionality the operator needs inside each process. The client application also manages the job workspace, where data is securely kept for the duration of the job.

MANAGING THE HIS DATABASE

BUILDING THE DATABASE

The main task during the first months of system operation was to load data from existing databases. Entering data into the HIS in a controlled manner was achieved by a workflow-controlled jobtype, where data went through the following steps of processing: registration, import, edit, quality control, validation, and commit. All user functionality is part of the ArcView GIS client application.

UPDATING THE DATABASE

The workflow for updating the HIS database is identical to that of building the database. Metadata that describes the cause of the modification is registered. Relevant data is checked out from the database during the input process, edited, quality checked, validated, and committed back to the database. Previous versions of the data go to the history tables. The multiscale transfer queue is updated, so that the modifications are forwarded to the scale databases.

MULTISCALE MANAGEMENT

Multiscale management is concerned with managing data at multiple scale levels in a controlled manner. The control mechanism for multiscale management is the transfer queue. It is a mechanism by which data that has already been entered into the main database automatically gets queued for transfer to the scale databases. The records in the transfer queue can be sorted and prioritized based on a number of different criteria, thus enabling a flexible workflow. All multiscale transfer is workflow controlled.

Data transfer is done manually, with tools to aid the operator in generalizing features when they are transferred from the main database to the scale databases. There are also tools to transfer modifications or deletion of features from the main database to the scale databases. The links to metadata are kept after the features have been transferred.

ENC PRODUCTION LINE

The ENC production line provides for the exporting of data for ENCs in the IHO S57v3 data format, making timely and accurate creations of ENCs possible. Key features of the ENC production line are:

- creation and maintenance of a database of ENC cell definitions (spatial extent and attributes);

- translation from the internal object model to the S57 version 3 object model;

- datum and projection conversions (if needed);

- electronic new cell and electronic revised cell production;

- automated extraction of modified data from the HIS database for revised cells;

- ENC metadata management; and

- custom-defined contents of the output ENC cells.

CONCLUSION

The HIS is a full-fledged system for hydrographic information management suitable for use by organizations such as National Hydrographic Offices. It works in a multiuser mode based on a Windows NT–UNIX client–server architecture. All data is stored in an RDBMS. Workflow-controlled input, edit, transfer, and export of data are all supported. During editing, data is securely managed in workspaces on both the server and the client side. The system also provides database support for cartographic product definition and is expandable to a full cartographic production system. ENC production is one of the key features of the system.

ACKNOWLEDGMENTS

The entire HIS development team contributed to the contents of this chapter, specifically Matti Palosuo and Petteri Soikkonen of Novo Meridian Ltd., and Kevin Howald, formerly of T-Kartor Sweden AB. The system was based on the design requirements of the Swedish and Finnish Maritime Administrations.

ABOUT THE AUTHORS

Katarina Johnsson, Ph.D., was a project manager and system developer for T-Kartor Sweden AB from 1996 to 2000. She is currently with Ericsson Technology Licensing in Lund, Sweden. Katarina earned a doctoral degree in photogrammetry and geoinformatics from the Royal Institute of Technology, Stockholm, Sweden, in 1994.

Veli-Matti Kiviranta is a GIS project manager for the Novo Group, one of Finland's largest information technology service providers. The Meridian subdivision of the company (www.meridian.fi) offers expertise in mobile and GIS-based Internet solutions, as well as information technology consulting, software, and product and operating services.

CONTACT THE AUTHORS

Katarina Johnsson
Karlavagen 3a
S- 291 54 Kristianstad
Sweden
katarina.johnsson@mbox301.swipnet.se

Veli-Matti Kiviranta
Novo Meridian Ltd.
Piispanportti 12 B
02200 Espoo
Finland
Telephone: +358 205 6686
Fax: +358 205 66 5110
veli-matti.kiviranta@novogroup.com

Chapter 9

Applications of GIS in the Search for the German U-559 Submarine: A Brief Case Study

MICHAEL COOPER, LAURA CRENSHAW, AND TRACIE PENMAN

VERIDIAN-MRJ TECHNOLOGY SOLUTIONS

FAIRFAX, VIRGINIA

EDWARD SAADE

THALES GEOSOLUTIONS (PACIFIC)

SAN DIEGO, CALIFORNIA

ABSTRACT ▶ Recent advances in deepwater search and salvage have put within reach shipwrecks once thought to be inaccessible. The well-publicized search for and location of the *Titanic* is one example. The most important step in successfully locating a wreck is defining the "search box," or large area in which the wreck is likely to be. Establishment of a search box calls for the pooling and reconciliation of sometimes contradictory information from multiple sources. For example, during World War II, British destroyers forced the German submarine U-559 to the surface and a boarding party recovered vital codebooks and equipment before the submarine sank. While the position of the wreck was presumed from the destroyer's location, subsequent analyses of surface currents, sea state, and prevailing winds at that time indicated a more northerly search box was probable. Other strategic devices went down with the submarine, but the wreck site was not investigated. Use of a GIS to plan wreck searches helps to organize and make use of the complicated sets of data, and ultimately, to define the search box. Once defined, the characteristics of the bottom, including depths, types and consistency of sediment, and other area features such as canyons or known debris, help the sidescan sonar operators plan specific search parameters such as two-direction, towfish altitude, and potential for false targets.

THE WORLD WAR II SETTING

As documented by Rohwer and Hummelchen (1992), in the fall of 1942, British destroyers supported by a Welsley aircraft forced the German submarine U-559 to the surface in the eastern Mediterranean Sea. British troops successfully boarded the submarine and managed to seize equipment, codebooks, and other documents vital to the decoding of German transmissions, but two men were lost when the submarine's captain scuttled his vessel. Though other strategic devices—most importantly, the M-4 deciphering machine— went down as well with the submarine, the wreck site was never investigated. Decades later, with the position of the wreck presumed from the destroyer's location, analyses of surface currents, sea state, and prevailing winds at the time of the boarding indicated a more northerly search box was probable. The following is a description of the GIS methodology used by Veridian-MRJ to organize and present data pertinent to the establishment of an exact search box. The study is of a hypothetical nature, similar to actual proprietary studies that are currently being performed by Veridian-MRJ for its customers.

GIS METHODOLOGY

Veridian-MRJ has built and maintains a number of global environmental and cultural databases at multiple scales in its GIS to support deepwater search and salvage studies (e.g., Cimino, Pruett, and Palmer 2000; Palmer and Pruett 2000). These databases can generally be used as is, but are often updated with higher-resolution site-specific information for the study area.

The first step in the analysis is to determine the general area in which the lost vessel (in this case the U-559) is likely located, and whether or not this is in an area claimed by a country. Using the Global Maritime Boundaries Database (GMBD) developed by Veridian-MRJ (Cimino et al. 2000, www.MaritimeBoundaries.com), the last known position of the lost ship is compared with the Territorial Sea, the Contiguous Zone, and the Exclusive Economic Zone claims of the Mediterranean Sea littoral nations. If the shipwreck is believed to lie within such an area, a permit is required from the claiming country in order for a search to commence. It was fortunate in this case that only Egypt had claims in the vicinity, and that the wreck site was outside of the claimed region. A permit for shipwreck search was therefore not required. Figure 1 shows the GMDB data relative to the proposed search area.

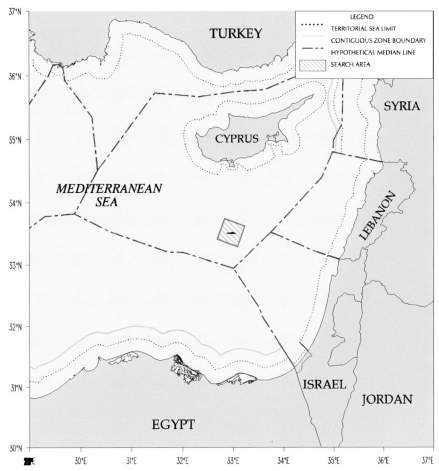

Figure 1. The search area for the U-559, shown relative to the claims and boundaries of littoral nations as maintained in Veridian's Global Maritime Boundaries Database.

Additional data retrieved from Veridian-MRJ's global GIS archives were then used to plan the search effort. Information relating to the ocean bottom such as bathymetry and seafloor sediments is shown in figure 2. The search area was defined based on the analysis of this data with two parallel goals: to allow the most efficient search methodology to be employed, and to minimize the impact on the ocean bottom during the search.

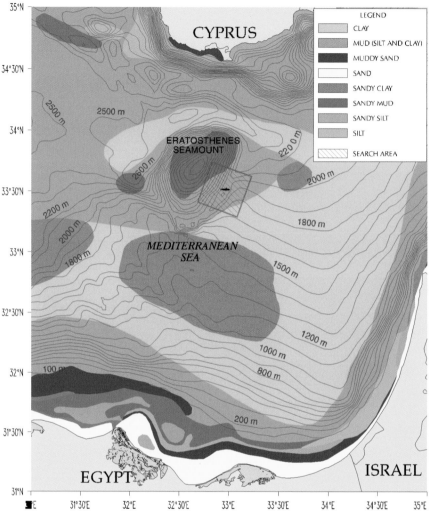

Figure 2. The search area for the U-559, superimposed on Veridian's global ocean sediment and global bathymetry databases.

This additional GIS data also provides a historical look at the weather that can be expected in the search area. Figure 3 shows a graphical representation of conditions for the period during which the hypothetical at-sea portion of the search for the U-559 took place. Time spent at sea for such a deepwater search is expensive (sometimes exceeding $40,000 per day for the lease of a vessel, the navigation system, and a sidescan sonar for mapping the seafloor), and a tool such as GIS that enables review of all data spatially can contribute to a successful and cost-effective search.

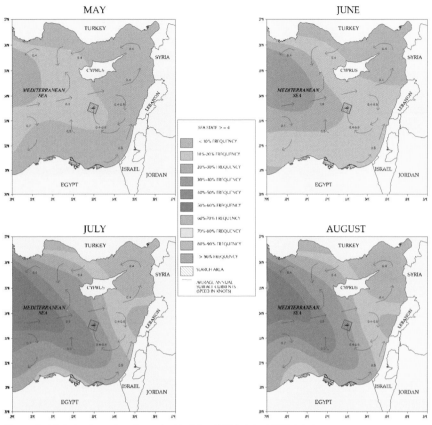

Figure 3. Historical weather from the Veridian-MRJ global GIS database.

Beyond the data sets depicted in figures 1, 2, and 3, many other data sets are used in planning an operation. In this instance, historical Notices to Mariners (NOTOM) chart correction and broadcast warnings from the National Imagery and Mapping Agency were reviewed to determine past activities in the vicinity of the proposed operation. This data can reveal whether previous surveys of the area have been performed and who performed them. These may turn out to be useful sources of new or higher-resolution data. Further, and perhaps more importantly from a safety standpoint, Veridian-MRJ's historical archive of NOTOM can indicate whether this area is being or has been used for military exercises or as an ocean dumping ground. Either usage could affect the safety of personnel and the viability of the operation.

Another use of GIS is shown in figure 4, where the positional data
of the towfish is entered into the GIS database to ensure that the
sidescan sonar vehicle has fully covered the search grid. This is to
ensure that the swaths of the vehicle have not missed the area where
the submarine might actually be located.

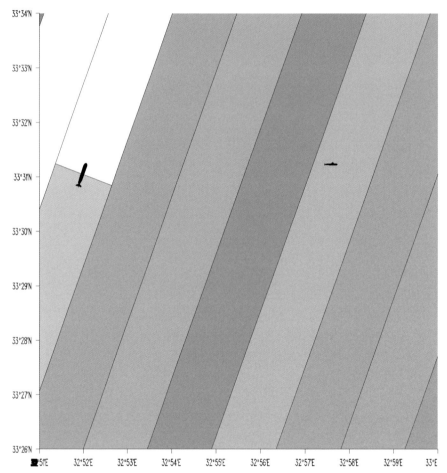

Figure 4. Marine chart created in a GIS showing the swaths of a sidescan sonar vehicle
(indicated by larger black symbol). Note the small horizontal black symbol near the center
indicating the probable location of the lost submarine.

ACOUSTIC MAPPING   In addition to making possible the location of sunken vessels such as
the U-559, recent developments in high-resolution acoustic sonar sys-
tems make possible an integrated systems approach to mapping seafloor features that can help researchers understand the geological

setting in which a shipwreck or artifact is located. These applications include mapping mobile sediments that may cover a wreck during investigation and salvage, and identifying the slump potential of regional sediment. This information can also aid sidescan sonar operators in planning specific search parameters such as two-direction, towfish altitude, and potential for false targets. For Veridian-MRJ in collaboration with Thales GeoSolutions (Pacific), the process often begins with high-frequency sidescan sonar imagery data collected and processed using TRITON ISIS software and mapped to support mosaic generation of the seafloor morphology (figure 5). Mosaics are generated using DELPH MAP software after all sonar data has been collected and integrated with differential GPS data (which tracks the surface vessel's position, and by extension, the position of the sidescan sonar vehicle). The mosaic functions as the base map for determining areas of interest and high importance. For example, figure 5 shows a 100-kHz acoustic mosaic exhibiting detailed rock texture suitable for geological analyses. Strong sonar backscatter in the mosaic is a probable indication of either steep slopes or coarse-grained sediment (i.e., cobbles and pebbles).

Figure 5. A 100-kHz acoustic image showing high relief and detailed rock texture. Dark reflections likely result from steep slope or coarse-grained sediment (cobbles and pebbles).

Designated areas are then mapped with a high-frequency multibeam echosounder to ensure the highest possible resolution and smallest individual beam-pattern footprint. Multiple swaths are collected across the feature to provide 100-percent coverage and a minimum of 10-percent overlap (figure 6). Using a remotely operated vehicle to deploy the multibeam echosounder, this method can be used in water as deep as 2,500 meters. The multibeam data is collected using

WinFrog Multibeam software, then cleaned, processed, and displayed in Caris Hydrographic Information Processing System (HIPS) software, which applies sun illumination results in shaded relief images and allows output in CAD format. The two data sets can be overlain in various formats for analysis and GIS support. The mapped regions yield an understanding of where future sediment slumps will likely occur and where the seafloor is currently mobile.

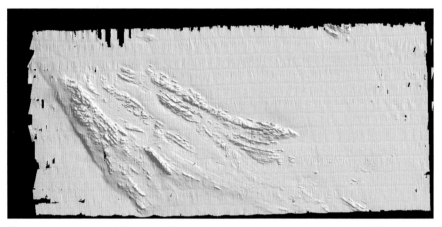

Figure 6. An example of digital multibeam bathymetric data collected using a 200-kHz multibeam echosounder in swaths to provide 100-percent coverage and 10-percent overlap. Caris-HIPS software was used to process this data with sun illumination applied to show shaded relief that can be compared with sidescan sonar data to produce optimum interpretation and seafloor characterization.

CONCLUSION

Ownership of, and rights to recover, antiquities and objects of value from the seafloor remain controversial issues. Knowledge of the extent of maritime boundaries, of a coastal state's jurisdiction over waters, and of the character of the seafloor are essential considerations for those seeking to legally salvage such objects. A GIS can greatly assist in the planning of an at-sea operation, reducing costs, avoiding hazardous consequences, and ensuring that an area has been thoroughly covered. GIS, in concert with combined geophysical and geological methodologies, can also constitute an integrated systems approach to mapping seafloor features, providing an efficient and economical way to image the shallow seafloor, and producing data that can be used to address many problems associated with deepwater search and salvage, as well as marine archaeology.

REFERENCES

Cimino, J. P., L. T. Pruett, and H. D. Palmer. 2000. Management of global maritime limits and boundaries using geographical information systems. *Integrated Coastal Zone Management* 1(3):91–97.

Palmer, P. and L. Pruett. 2000. GIS applications to maritime boundary delimitation. In *Marine and Coastal Geographic Information Systems*, Wright, D. J., and D. J. Bartlett, eds. London: Taylor & Francis.

Rohwer, J., and G. Hummelchen. 1992. *Chronology of the War at Sea, 1939–1945.* Annapolis, Maryland: Naval Institute Press.

ABOUT THE AUTHORS

Michael Cooper has been involved with search and salvage projects for over twenty years, including the space shuttle *Challenger*, a South Africa Airways 747 lost off Mauritius, a United Airlines 747 cargo door lost off Honolulu, Hawaii, and numerous shipwrecks. He has also conducted surveys for and installed telecommunication cable systems, and assisted governments in locating and recovering lost items of interest.

Laura (Lu) Crenshaw has more than ten years of experience in surveying, mapping, data capture, and analysis, and has been with Veridian-MRJ Technology Solutions for five years.

Tracie Penman is a GIS and data analyst for Veridian-MRJ Technology Solutions and has been involved in the GIS industry for more than thirteen years, specializing in ArcPlot™ and hard-copy output. She searches for the latest data in marine environments and is a member of IAMSLIC, a marine science library dedicated to the exchange of digital marine data.

Edward Saade is the general manager for Thales GeoSolutions (Pacific) in San Diego, California. He holds a bachelor of science degree in geology from the University of California, Santa Barbara, completed graduate studies and research in marine geophysics at the University of Hawaii, and is a California Registered Geophysicist. He was involved in the search and salvage of the TWA 800 crash off Long Island.

CONTACT THE AUTHORS

Michael Cooper
Michael.Cooper@veridian.com

Laura Crenshaw
laura@veridian.com

Tracie Penman
MaritimeBoundaries@veridian.com

Veridian-MRJ Technology Solutions
Attn: Maritime Boundaries
MS 3A
10560 Arrowhead Drive
Fairfax, VA 22030
Telephone: (703) 277-1212
Fax: (703) 385-4637
www.MaritimeBoundaries.com

Edward Saade
Thales GeoSolutions (Pacific)
3738 Ruffin Road
San Diego, CA 92123
Telephone: (858) 292-8922
Fax: (858) 292-5308
Edward.Saade@thales-geosolutions.com

Chapter 10

From Long Ago to Real Time: Collecting and Accessing Oceanographic Data at the Woods Hole Oceanographic Institution

ROGER GOLDSMITH

WOODS HOLE OCEANOGRAPHIC INSTITUTION

WOODS HOLE, MASSACHUSETTS

ABSTRACT ▶ Oceanographers have been developing specialized software for the analysis and display of their data for many years. As the GIS field has matured with the development, for example, of new Web-oriented products, oceanographers have begun to recognize how much desktop products such as ArcView GIS can contribute to their research. This chapter describes three prototype, Web-based applications that were developed to make researchers at the Woods Hole Oceanographic Institution aware of the capabilities of current GIS technology: (1) SEDCORE2000 for geological oceanographers, an interface for viewing, querying, and analyzing information on sediment cores, rocks, and other marine geological artifacts recovered from the ocean floor; (2) a system for physical oceanographers that allows researchers to view, query, and download data from current meters and moored instruments for tracking ocean currents; and (3) also for physical oceanographers, an application in support of the Atlantic Circulation and Climate Experiment for monitoring and broadcasting the progress of ocean floats that help to characterize the general circulation and water mass structure of the North Atlantic. Also discussed are some problems that need to be addressed in order to achieve a more effective integration of GIS and oceanography.

INTRODUCTION

The collection of oceanographic data began long before the era of the electronic computer, relational databases, and modern geographic information systems (GIS). Consequently, some unique data storage formats and analytic tools have been developed to address the specific needs of oceanographic research. Oceanography is still very much an academically oriented research "industry," and a lot of the work is typically done in small, scattered laboratories. The oceans cover a large area, and therefore the data sampling, although deep, is sparsely distributed in breadth (area). Often sampling has a three-dimensional depth component and therefore may have been repetitively sampled at one geographic location. However, this same multisource, uneven distribution is ideally suited to the geographic display powers of GIS, which allows data of various types and scales to coexist in the same system. As research in oceanography has become increasingly interdisciplinary, researchers want access to current as well as historical data. This chapter summarizes three projects that are being developed to exploit the capabilities of GIS, and addresses some of the needs for summarizing and sharing data.

SEDCORE 2000

The Seafloor Samples Laboratory at the Woods Hole Oceanographic Institution (WHOI) maintains a collection of more than thirteen thousand samples of sediment cores, rocks, and other marine geological artifacts recovered from the bottom of the ocean (figure 1). Sampling of the seafloor goes back to the days when sailors lowered wax-coated stones and continued with the use of tallow-filled "leads." With the advent of drilling technology, researchers can now capture not only the surface of the seafloor, but also long, stratified samples tens of meters in length. Because the sediment cores are comprised in large part of particles settling down from the sea surface above, the skeletal remains of small plankton can be used to reconstruct climatic changes from the recent to the distant past. Detailed records of many other types of events, from volcanic eruptions to sea level variations, are found in the stratigraphy of deep ocean sediments.

Figure 1. Sections of seafloor sediment cores are analyzed by hand. The cores are stored for future analysis and the descriptions are now accessible through an interactive GIS server (courtesy of J. Broda, WHOI).

The organized analysis and description of WHOI seafloor samples began in the 1960s, an era that predated the widespread use of computers and storage of data in digital form. The cores were described using visual and microscopic techniques. The first resulted in a lithological log, a subjective description of the various component segments of the core, which nevertheless makes use of objective classification schemes to illustrate core content. As can be seen from figure 2, visual impressions of color, texture, and the relationship between various features in the core were noted. The second technique involved the preparation and microscopic analysis of tiny sediment subsamples taken at specific intervals throughout the core. The abundance of standard constituents was estimated, then used to enhance the visual description (figure 3). To date, eight comprehensive volumes of the hand-generated descriptions have been published and distributed globally. They are still the primary reference resource of the curator and many outside users of the collection. (Mills and Broda 1993).

1221

VISUAL CORE DESCRIPTION Page 1 of 1

Ship CHN Cruise 7 Leg ~ Sta. SC6 Core No. SC6
Total Length 135 cm. Lat. 36°12.5'N Long. 23°21.9'E Depth 1194 m.unc
Core condition Poor: cracked dried, separated. Date Described 21MAY74 by T. Broda.
Physiographic location E. MEDITERRANEAN SEA.

Lithologic Log

Detailed Description

0-19
CALC OOZE
2.5Y 7/2 lt gray
G
19-24
CALC OOZE
2.5Y 6/2 lt brownish gray
with abunt. black flecks
S inclined 10°
24-111
CALC OOZE
2.5Y 7/2 lt gray
scattered black flecks and small white mottles
scattered forams
G
111-121
CALC OOZE
5Y 6/2 lt olive gray
abunt. white flecks
S
121-135
CALC OOZE
2.5Y 7/2 lt gray
common mottling, 5Y 6/2 lt olive gray
few forams
end of core

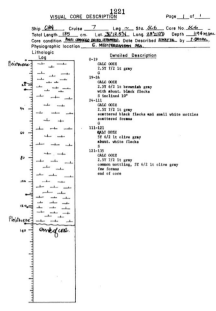

Figure 2. An example of an original hand-generated lithologic description of a core taken from the seafloor. This is a general descriptive analysis of identifiable sections within the core.

1222

SMEAR SLIDE DESCRIPTIONS - W.H.O.I. SEDIMENT CORES

Ship: Chain Core No. 6 SC
Expedition 7 Station No. 6
Leg No. Total Core Length 135 cm

LEVEL	SEDIMENT TYPE	ESTIMATED ABUNDANCES (%)													
		Inorganic Material					Biogenous Material								
		Silt & Sand					Calcareous					Siliceous			
		Detrital grains	Micronodules	Zeolites	Volcanic shards	Clay	Forams	Nannofossils	Pteropods	Discoasters	Others	Diatoms	Radiolaria	Sponges
2 cm	calc ooze	3		tr		43	3	43	tr		7			1
70 cm	calc ooze	2			3	44	2	44	tr		5			tr
114 cm	calc ooze		6			60	3	20			10			1
134 cm	calc ooze	3				40	2	48			7			tr

Figure 3. An example of an original hand-generated analysis of seafloor core. This is a more numerical description of the constituents found at selected levels throughout the core.

Several years ago a project was initiated to automate the generation of the descriptive diagrams. While the analysis still depends on the very human visual inspection of the sample, the data is now entered into a digital database. Software programs such as SEDCORE 2000 allow for the generation of the graphics; while they may lack the distinctive human touch, they are much easier to maintain, modify, and reproduce (figures 4 and 5). Most importantly, the information about the collection is easier to disseminate.

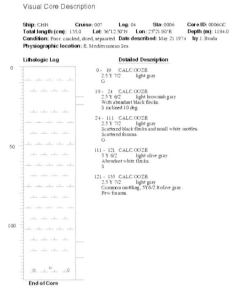

Figure 4. Lithological description generated by the desktop version of SEDCORE 2000 software (not yet GIS Web-enabled) using descriptive information extracted from the database.

Smear Slide Description

Ship: CHN   **Cruise:** 007   **Leg:** 04   **Sta:** 0006   **Core ID:** 0006GC
Total length (cm): 135.0   **Lat:** 36°12.50'N   **Lon:** 23°21.90'E   **Depth (m):** 1194.0
Condition: Poor: cracked, dried, separated   **Date described:** May 21 1974   **by** J. Broda
Physiographic location: E. Mediterranean Sea

LEVEL	SEDIMENT TYPE	ESTIMATED ABUNDANCES (%)												
		Inorganic Material					Biogenous Material						
		Silt & Sand					Calcareous					Siliceous		
		Detrital grains	Micronodules	Zeolites	Volcanic shards	Clay	Forams	Nannofossils	Pteropods	Discoasters	Others	Diatoms	Radiolaria	Sponges
2	CALC OOZE	3		TR		43	3	43	TR		7			1
70	CALC OOZE	2			3	44	2	44	TR		5			TR
114	CALC OOZE		6			60	3	20			10			1
134	CALC OOZE	3				40	2	48			7			TR

Figure 5. Slide analysis generated by the desktop version of SEDCORE 2000 software (not yet GIS Web-enabled) using numerical information extracted from the database.

While this mode of analysis and data entry was still in its infancy, it highlighted the need for better access to the collection, both for internal use, and increasingly for use by other investigators. While individual ship cruises were summarized in each published volume, there was no comprehensive map of all the locations. This seemed like an ideal task for a GIS. The historic forerunner was the MUDDIE system, a punched-card data format that could be searched by a FORTRAN language program. This produced a list of sites selected on the basis of location or a sampling event.

The data from that system forms the foundation for the initial population of the new database. Information is extracted from the relational database as dBASE-III or text-format files and input to the GIS. While it is possible to link many of today's GISs directly to the database, time was not an especially critical factor in this application. There were logistical reasons, as well, for keeping the database and the GIS uncoupled during this development phase—not the least of which was the fact that collection managers were stationed abroad for several years.

After several iterations, the basic Web-served version of the seafloor samples collections interface (SEDCORE 2000 based on ArcView Internet Map Server) appears as shown in figure 6. At the general information level, users can search for samples much like they did with the old MUDDIE system, except that the interface is now much more visual and geographically based. This is a primary strength of a GIS. A second advantage is the ability to separate or combine information from several categories or themes. The Seafloor Samples Laboratory has much data associated with cores and dredges or grabs. The information is similar for each type, but not identical. And yet in SEDCORE 2000 it is integrated into a composite view of the entire structure, both of the collection and ostensibly the seafloor itself. The seafloor depth has been added as a cue, putting the cores into some semblance of its actual three-dimensional context. Conceivably, layers containing the sediment thickness or type could be developed and overlain to provide additional information.

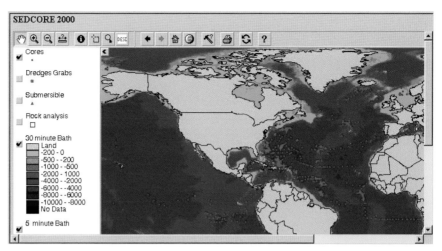

Figure 6. Initial SEDCORE 2000 Web interface, showing the sample locations of the cores at WHOI.

At this entry-level point of inquiry, details about the core site (in space), the sampling (in time, equipment, and technique) and at least an overview of the description are available. Almost all of this information can be queried to further refine the selections to match the researcher's interests. The user might, for instance, like to obtain more information about the sample itself. What is the composition of constituents (slide analysis) at a given level? What kind of distribution of rock types from a dredge are there? What is the lithologic description for a selected segment of a core? The capabilities of the database and GIS will easily handle this type of detailed data extraction. One problem persists, common to many GIS projects: collecting and preparing nondigital data. Only the new analyses are being done digitally. And while there are plans to transfer the archived manuscript volumes to digital format, it remains to be done. How can such data be made available now?

The solution is quite straightforward: make the original descriptions available on the Web. Each page was scanned, stored in Adobe® Portable Document Format (PDF), and cross-referenced to its entry in the database. The GIS was modified to provide an image reference in the summary list (figure 7). Having narrowed the search, the user may now summon up the detailed lithological description and slide analysis. While this does not support the actual use of the detailed information in digital text searches, it does provide the researcher with all the data relevant to the collection.

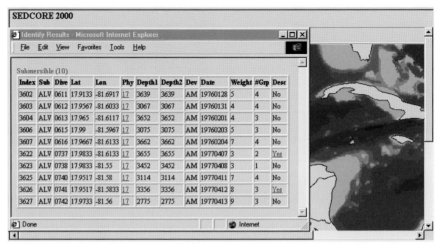

Figure 7. Second SEDCORE 2000 Web interface, showing an image reference with database summary list.

With the success of the SEDCORE 2000, two additional projects at WHOI were developed incorporating GIS as a central component. This chapter will concentrate on the Web-based accessibility portion of the applications, although they are also being used for spatial analysis. The purpose is simply to provide researchers with the resources they need to do their research. Most GIS veterans will agree that data collection and organization is the major part of any GIS effort. Oceanography is no different. Scientists need a way to inventory who did what, when. (Or perhaps where nothing has been done; most of the ocean remains effectively unmeasured.) They would like to view the historical record not as a group of individual projects but as a collection of data.

TRACKING OCEAN CURRENTS

In the scheme of things, the geology of sediments does not usually change much over short periods of time. Currents, on the other hand, vary at all scales, depths, and times. Measurements of ocean currents by scientists at WHOI go back about forty years. They involve mooring an instrument in the ocean and continuously measuring the direction and speed of the water mass, as well as additional properties such as temperature or conductivity. The purpose of WHOI's second project was twofold: (1) to collect the data in a common repository with common formats; and (2) to allow researchers to view and query data availability, then download it as desired. The parameters being measured were not isolated, single-occurrence events, but rather the dynamic, time-varying properties of the ocean at a specific geographic location. Time-dependent data set off alarm bells in the minds of many GIS users, and this application had to confront the problem.

Figure 8 shows the initial cut at an ArcView GIS project. Superficially it is just a geodata inventory problem. The global view of the WHOI moorings seems to contain a relatively manageable number of sites. However, closer inspection of a site reveals clusters of moorings set for specific experiments. In fact, there are about 750 unique geographic locations in the WHOI subset. And as figure 9 illustrates, a single deployment may have several instruments associated with it. Right away, one is measuring in three dimensions. When using the traditional GIS techniques to obtain information at a single geographic location, the user is confronted with possibly tens of "hits." On large regional or global charts the number could be hundreds. The WHOI collection has more than 2,700 instruments, considerably less than 10 percent of the total number of instruments. When the sampling rates and number of parameters are taken into account, this set alone contains over three hundred million observations or readings. Storing these as single, geographically referenced values might prove useful for generating climate snapshots. At this stage of computational technology, however, it might make retrieval of time series data cumbersome.

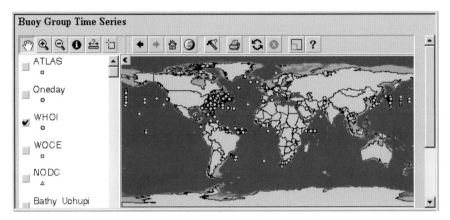

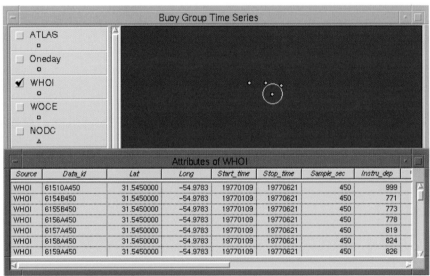

Figure 8. The initial Buoy Group Time Series Web presentation showing (top) the locations where WHOI current meters collected data and (bottom) an enlargement of an area showing a few WHOI time series data locations and their associated attributes.

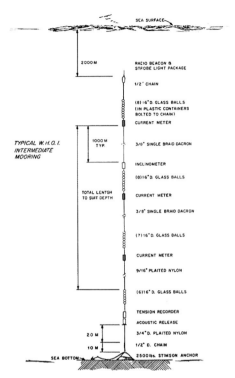

Figure 9. Schematic drawing showing typical oceanographic moored instrument deployment.

As stated earlier, the objective of the project was to put data online so that researchers could download it. In the past, researchers inquiring about the availability of data would be referred to the individual who knew the legacy, and be sent cards, tapes, discs, CD–ROMs, or electronic mail, as the technology progressed. At about the same moment, it was realized that no one would be able to grasp the multidimensionality of the collection without proper metadata. Because location is still the usual first cut in making a selection, a GIS is again a logical choice to provide the method for accessing the metadata. At about the time the design was taking place, ESRI was also unveiling its network-accessible extension to ArcView GIS, Internet Map Server software, which would be especially appropriate for serving the metadata.

The metadata had to provide a portal into the data itself, so the design of its structure was especially important. Certain features were obvious. Where was the data located (latitude, longitude, height above bottom, depth)? When was it measured and how often (year, month, day, hour, minute, second for both the start and end of the time series, as well as the sampling rate)? What was being measured (parameters) and where was it being stored (file access)? It was also necessary to know something about the time series data itself. Was it a "normal" time at the site? Was the instrument reliable? Some of these questions could be answered, or at least so it was initially thought, by providing a statistical summary of key parameters, including selected averages and extremes.

At this point one of the problems of dealing with historical data became apparent. Oceanographic data collection was, and in many cases still is, dependent on a power source. Traditionally this has been batteries, and as the technology has improved the data length has been extended, or the sampling rate increased. Parameters might be measured as infrequently as once a day or as often as every second. Because of limited power source lifetimes, there were often several deployments at the same location, adding the element of time-dependent measurement capability. As instruments improved, new sensors were added and different parameters were measured. Add to this the fact that data was collected from several sources, from various principal investigators, using a variety of formats and conventions for situations such as missing data, and one gets an idea of the problems confronting the data quality manager. (And, of course, this was being done in a limited-resource academic research environment.)

What was really needed was a way to quickly scan the time series itself. A decision was made to try the graphics formats commonly used to display temperature and currents. The entire series, filtered or subsampled as necessary or appropriate, could easily be pregenerated and put on a single chart. Users could then, after narrowing their choice, display the chart and immediately see any trends, discontinuities, or irregularities in the sampling. The data files would be stored in both ASCII and netCDF formats so that the user could look at actual values before downloading a selected file. Figure 10 illustrates the results of the Identify tool within an ArcView Internet Map Server application. Note that many hits are returned for a single location. The user can determine which instruments meet other criteria and then view or download specific files (figure 11). For the more general type of query, such as those instruments meeting some two-dimensional spatial criteria, the number might be extensive; the selection box tool was therefore modified to return only a list of those instruments in an area. The list could then be saved and later used to download a larger number of files.

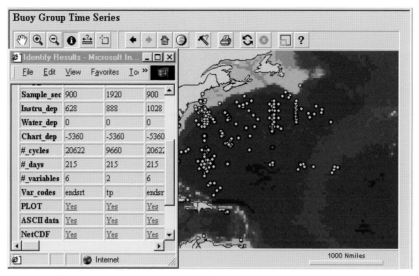

Figure 10. Selecting a location for further information can, depending on the scale, result in the identification of several instruments in close proximity. Even a single mooring may have many instruments located at various depths.

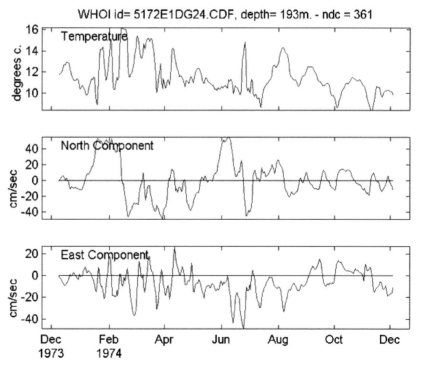

Figure 11. A typical instrument parameter plot, pregenerated and made Web-accessible for users wishing to look at the time series characteristics prior to downloading the data files.

Another important design consideration was documentation, which should include not only the notes and comments associated with the data collection, but at least some instructional reference for using the interactive retrieval system. This was implemented as an "Info" page, accessible both at the login and from a button on the map page. The documentation gives a more detailed explanation of the various codes and abbreviations used in the metadata, as well as providing some context and instruction for using the interface. Notes for specific instruments, which may range from nothing to lengthy explanations, were made available in a manner analogous to the plots (i.e., through the detailed identification results).

It was previously mentioned that some statistical measures were made for each instrument, and it might seem reasonable that these could be used in conjunction with a query button to refine selections. But in the current implementation no statistics are included. One of the lessons reinforced in the initial deployment of this GIS application concerns the accuracy of historical data. There are inevitably some errors but there are many other issues. It is emphasized again that the observations have been collected from many scientists, from several countries, over many years. Has the precision changed over the years? Have techniques changed, perhaps introducing a systematic difference? Are these all the same units? Have codes for things such as "missing values" in a time series differed? While these factors are readily discernible for a specific instrument, they are considerably more difficult to deal with in preparing the metadata. And, as was the case here, if some of the statistical results were questionable, then their use in queries would compromise any selections obtained. This has introduced a much more thought-provoking question. Who maintains historical data? If a value, such as latitude, is apparently in the wrong hemisphere, or is incorrect, is it proper to fix the data? Is there a more serious underlying problem? Is it even expeditious? Or is it better to adjust the extracted metadata for which we do have some responsibility and leave the historical record as collected? These are issues still under consideration.

Moored buoy platforms were developed, in part, so that the instrument could be found later. GPS-enabled advances in acoustic technology (e.g., listening stations, then listening floats) were developed that enabled the tracking of free-floating buoys. Organized experiments have been taking place since the 1970s, initially with surface floats, then subsurface floats, and now with instruments that can repeatedly dive and return to the surface. The purpose is still to track the movement of parcels of water and measure the temporal change of properties such as temperature and salinity. Whereas the initial instruments required the researcher to recover the listening station or float to collect the data, instruments are now able to transmit the data to a new form of listening station, the orbiting satellite. The message can then be sent as electronic mail to the researcher's desktop computer. In other words, it is now possible to track and acquire data in real time.

Drifting buoys, both surface and subsurface, have been used for more than thirty years to track the ocean's currents. Figure 12 shows a subset of surface drifters and subsurface floats. The subsurface floats, put at varying depths down to 4,000 meters, are particularly interesting because they "feel" a lot more ground than is seen in a normal view of the sea surface. Figure 13 is an example of the impediments that a float at 700- or 1,500-meter depth might encounter in the North Atlantic. Even when not running aground, some seemingly anomalous behavior can be explained with the introduction of bathymetry (figure 14).

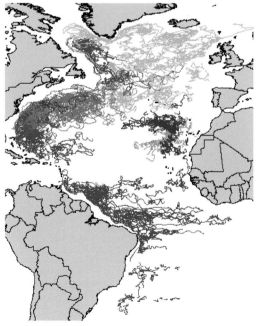

Figure 12. The trajectories of surface drifters and subsurface floats can be easily displayed using a variety of symbolization.

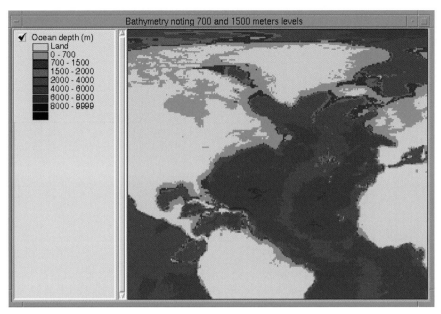

Figure 13. Subsurface floats operating at 700- and 1,500-meter depths have a significantly restricted area in which they can operate without running aground.

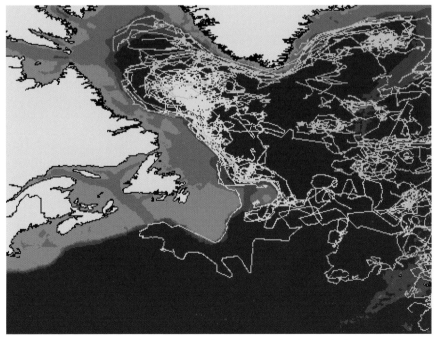

Figure 14. The introduction of bathymetry to the trajectory plots helps clarify some of the processes and seemingly anomalous behavior.

A third prototype GIS application was developed for the Atlantic Circulation and Climate Experiment (ACCE). This involved the launching of PALACE floats in the North Atlantic Ocean (figure 15) to track its circulation. In general, the instrument is ballasted to be neutrally buoyant at some depth and to float in a water mass for about ten days. It then rises to the surface, measuring various properties of the water column on the way up. Remaining on the surface for a short time, it transmits its position and data to a satellite, then repeats the cycle by sinking to a programmed depth.

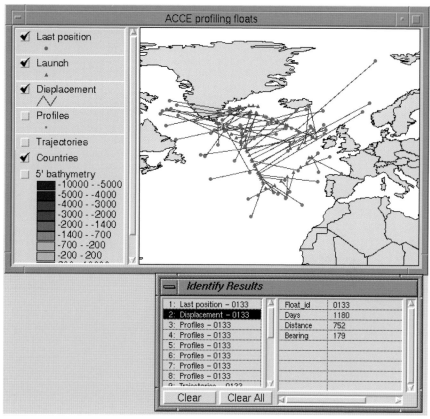

Figure 15. A quick overview of the float histories can be obtained by viewing the launch positions, the current positions, and the new displacement vector.

GIS proved to be a useful tool for monitoring and broadcasting the progress of the floats. The overview being served on the Web allows scientists to view the current position of any float and determine the trajectory as inferred from its surface locations (figure 16). To simplify this "spaghetti bowl" (and this is only a relatively small experiment), launch sites of selected buoys can be shown, as well as net displacements. The locations where the instrument profiled the water column on its way to or from the surface may also be displayed (figure 17). This profile data is currently not available in real time, as the studies are still in progress.

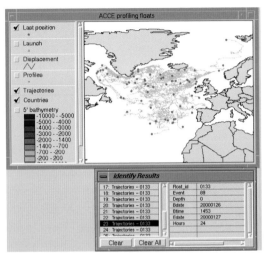

Figure 16. The trajectories of the floats can be inferred from where they rose to the surface. These trajectories have been broken down into discrete cycle segments.

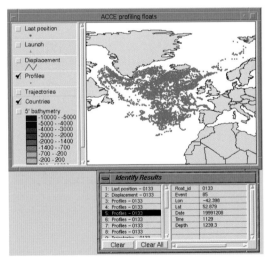

Figure 17. Properties of the water column were measured by the float on its way to or from the surface. The metadata provided by the identification can be used to retrieve data for a specific profile.

With regional and global profiles, scientists can obtain a snapshot of the three-dimensional structure of the water mass. The three-dimensional aspect of the environment is a very important component in the analysis of most earth science data. Standard, single-valued functions of x and y alone are no longer sufficient; volumetric analysis that can be performed with software packages such as Advanced Visualization System (AVS), Matlab, or Surfer are now necessary. While some GIS packages offer a limited three-dimensional analytical capability, most of the analyses of the WHOI projects are still done using non-GIS software. As an example of the increasing capabilities of GIS, however, the profile data has been used to generate temperature fields (figure 18) and the depth of the thermocline (figure 19). Some GIS packages, such as ArcView software's Tracking Analyst extension, provide a good start at addressing the time-varying position problem, and allow some depiction of float trajectories (figure 20).

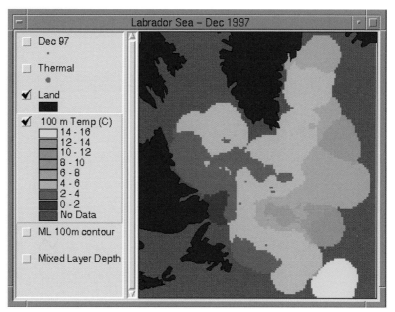

Figure 18. Desktop GIS packages are increasingly adding more sophisticated tools, such as allowing the contouring of the water temperature at 100-meter depth in the Labrador Sea.

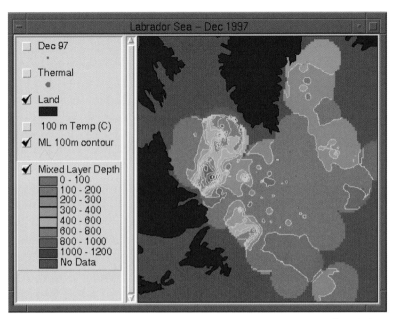

Figure 19. A contour analysis of the mixed layer depth in the Labrador Sea during December 1997.

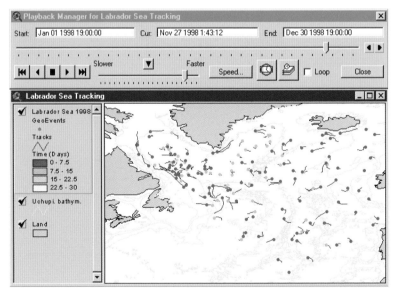

Figure 20. The ArcView Tracking Analyst extension allows users to view time-varying features.

Information learned from developing this GIS application will be useful within a larger physical oceanographic experiment called Argo. Argo will include the global deployment of more than three thousand floats, and will bring into consideration new operational factors such as weather conditions and jurisdictional boundaries (figure 21). All of these components lend themselves to the theme or layer concept of GIS.

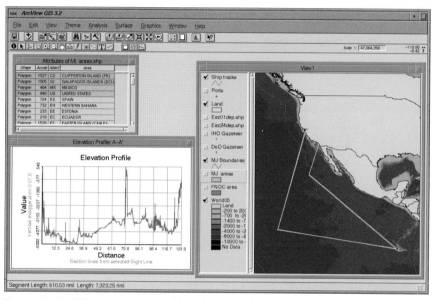

Figure 21. The GIS capability of combining layers such as bathymetry, service areas (ports), and jurisdictional claims make it a useful tool for mission planning and operations support.

CONCLUSION   GIS technology now allows for the synthesis of oceanographic data collections, particularly the geological and physical oceanographic applications described in this chapter. Interdisciplinary research requires the assimilation of data from many different sources or themes, and this is one of the primary strengths of GIS. The use of metadata, too, allows the incorporation of multidimensional data such as depth and time where it might otherwise not be practical. And finally, the use of historical data is important, as oceanographers are just starting to investigate long-term, decadal time scales in climate fluctuations. Legacy data, often stored in a variety of visual formats, can be made more widely available when used in conjunction with metadata summations and modern GIS.

ACKNOWLEDGMENTS   The author would like to thank the following WHOI staff for their
contributions of data and time: Mr. Jim Broda, Dr. William Curry,
Ms. Ruth Goldsmith, Dr. Nelson Hogg, Dr. W. Breckner Owens,
Rear Admiral Richard Pittenger, USN (Ret.), and Ms. Susan Tarbell.

REFERENCE   Mills, P. B., and J. E. Broda, eds. 1993. Descriptions of WHOI Sediment Cores, Volume 8. *WHOI Technical Report* 93–19. Woods Hole, Massachusetts: Woods Hole Oceanographic Institution.

RELATED RESOURCES/
FURTHER READING   Unidata. DODS: Distributed Oceanographic Data System. Electronic journal: www.unidata.ucar.edu/packages/dods. Boulder, Colorado: University Corporation for Atmospheric Research.

Unidata. NetCDF: Network Common Data Form. Electronic journal: www.unidata.ucar.edu/packages/netcdf. Boulder, Colorado: University Corporation for Atmospheric Research.

Wessel, P., and W. H. F. Smith. 1995. New version of Generic Mapping Tools released. Eos, *Transactions of the American Geophysical Union* 76(33):329.

Wright, D. J., R. Wood, and B. Sylvander. 1998. ArcGMT: A suite of tools for conversion between ARC/INFO and Generic Mapping Tools (GMT). *Computers and Geosciences* 24(8):737–744.

ABOUT THE AUTHOR   **Roger Goldsmith** is an information systems specialist at the Woods Hole Oceanographic Institution, and has been at WHOI since 1972. He has combined his background in meteorology, oceanography, and geophysics with an interest in geography, mapping, and visualization to develop improved techniques for analyzing and presenting data.

CONTACT THE AUTHOR   Roger Goldsmith
Woods Hole Oceanographic Institution
Computer and Information Services
161A Clark, Mailstop 46
Woods Hole, MA 02543
Telephone: (508) 289-2770
Fax: (508) 457-2034
rgoldsmith@whoi.edu

Chapter 11

Using the Internet to Manage Geospatial Submarine Cable Data

DAVID CASWELL, BILL GILMOUR, AND DAVID MILLAR

THALES GEOSOLUTIONS, LTD.

SAN DIEGO, CALIFORNIA

ABSTRACT

Improvements in computer hardware and software systems have led to significant opportunities to improve the manner in which submarine cable data is managed and distributed. This is particularly important given the increasing number of cable systems being built, and subsequent needs for maintenance and protection of these systems. The fast-track conditions in which these systems are now engineered and installed also means that data must be exchanged and analyzed faster than ever before without any compromise to data quality or accuracy. These increased demands mean that, in many cases, traditional methods break down and new methods must be used to meet the requirements of very aggressive schedules.

This chapter describes how Thales GeoSolutions has used recent advancements in data management, geographic information systems (GIS), and the Internet to develop a system that significantly improves the manner in which route position list (RPL) and geospatial submarine cable data in general is managed and distributed. The system focuses on the fundamental architecture of cable data rather than on standard file formats, thus enabling implementation of cable data in an Internet-accessed relational database. The result is an Internet-based delivery system that provides users with immediate and instantaneous access to various sources of cable data. This development of "GIS through the Web" will promote as well the use of distributed databases and allow remote interaction with these databases.

INTRODUCTION

The quantity and complexity of submarine cable installation data has increased dramatically over the past five years. This is partly due to the significant increase in the number of cable systems that are being installed, but is also the result of the digital survey technologies that are now available for route and "as-laid" data acquisition. The increased volume and complexity of this data has presented significant challenges to the organizations involved, which are all required to interpret and distribute cable-related data. These organizations include marine survey companies, cable installers, cable owners, permitting agencies, government regulators, insurers, investors, and cable maintenance authorities. Even relatively simple cable data, such as route position list data—coordinates that define the horizontal geometry of a cable route—are handled with difficulty by many submarine cable organizations in various ad hoc or proprietary data formats. Thales GeoSolutions (formerly Racal Pelagos) has been a provider of submarine cable services and software for more than ten years, and has recently made several advances in the management of submarine cable data, based on GIS and Internet technologies. By using commercial off-the-shelf software, combined with custom-developed applications, Thales GeoSolutions has created an Internet-based, cable-specific data management and data distribution system. The company has successfully used this system to support operations for both in-house and external customers.

CURRENT PRACTICES

A number of recent developments in the submarine cable industry have conspired to render the handling and use of cable data increasingly difficult. These developments include operational problems associated with the sheer number of new cable projects, but there are several significant technical issues that need to be addressed as well.

As the number of submarine cable installations has increased, the allotted schedule for a typical installation has been compressed. The time between concept and the start of cable-laying operations can now be as little as nine months, and data-oriented deliverables such as desktop studies, cable route surveys, route design, and cable engineering are now completed in weeks or months, instead of months and years. The difficulties presented by these compressed timelines are compounded by an increase in the number of multipartner or consortium projects, involving dozens of client partners, all with an interest in the installation. There is also an increasing need to distribute cable-related data to various third parties, such as permitting agencies, governments, resource management groups, insurance organizations, and investment groups. As a result, submarine cable data has a much wider audience than it had even five years ago, and is becoming more important to a greater diversity of stakeholders every year. These changes in how submarine cable data is used and distributed have occurred in the context of the rapid development of the Internet, which has promoted instantaneous, on-demand access to information from anywhere in the world. The business, social, and operational environment in which submarine cable information is used is therefore much more demanding than it was in the past.

Just as the demand for enhanced data use and data distribution has increased, improved sensor and instrumentation packages now deliver more data during acquisition. New technologies such as the differential Global Positioning System (DGPS), multibeam echosounders, and digital sidescan sonars, combined with highly instrumented cable engines, ploughs, and remotely operated vehicles (ROVs), have increased the quantity of available submarine cable information by many orders of magnitude. This, in turn, has placed strains on the historical methods and systems used to handle and process the data. Related problems include the emergence of a wide range of proprietary and incompatible data formats, difficulties in tracking versioning and data-editing operations through time (especially important in multiversioned cable route designs), and challenges associated with ensuring the quality and security of data and data products.

OBJECTIVES
The goals of a generally applied technological solution to the problems of data management and distribution that now face the submarine cable industry are several. A key objective is to make the information that can be extracted from the large volumes of available raw data as useful as possible. This objective is applicable to all stages of the submarine cable operations, from extracting the most cost-effective route via desktop study or route survey data, to extracting valuable maintenance information from raw as-laid data. It is ultimately the value of this extracted information that determines the value of the entire data acquisition and processing effort.

A secondary goal is to find a way to distribute data and information to stakeholders located around the world. This must be accomplished in the context of accelerated schedules and reduced timeframes, without compromising the quality and integrity of the data. The users of submarine cable information are now demanding almost instantaneous access to data worldwide. There is also a growing need to provide these users with a means to interact with the data through feedback, interpretation, and modification. The solution must therefore provide for efficient management and redistribution of such interactive contributions.

The final goal is to manage data in a way that promotes confidence in the data products. As the value of submarine cable infrastructure increases, the consequences of an error or damage to information describing that infrastructure becomes more severe. Consequently, mechanisms for ensuring the quality and security of data are of paramount importance in any proposed solution.

Reducing the costs associated with submarine cable data acquisition and processing is another important goal of any solution. This goal can be achieved in several ways, including the intelligent and licensed reuse of proprietary and public domain data. Costs could also be reduced through improved efficiencies in data-processing and data distribution mechanisms, as well as the use of relatively inexpensive public domain data to replace or enhance expensive survey-acquired data.

THE THALES GEOSOLUTIONS APPROACH

Thales GeoSolutions has been applying new technology to submarine cable data management since the late 1980s. The company pioneered the development of cable-specific integrated navigation systems for installation operations and the on-board documentation of as-laid cable data. More recently, submarine cable data management systems were developed and delivered to several major cable companies. The company has also created and is operating a comprehensive cable data distribution site on the Web. These initiatives, combined with the development of a submarine cable data model, make it possible for Thales GeoSolutions to address the major data management and distribution issues now facing the submarine cable industry.

The Thales GeoSolutions approach to cable data management has been centered on the use of four key software technologies. At the core of the system is a data architecture, or schema, that accurately models cable data in a consistent and logically correct format. This data architecture determines what queries and operations are possible for the overall system and is therefore a critical component of the overall system. The four software technologies that combine in this solution are summarized in the following sections.

GEOGRAPHIC INFORMATION SYSTEM TECHNOLOGIES

Commercial GIS packages incorporate a range of data management functions, including the graphical presentation of data and metadata, spatial analysis functions, and charting. Cable and cable-route data is inherently spatial, and most effectively managed and used as geographic data. New products from ESRI also provide mechanisms for sharing cable data within a local workgroup or with remotely connected colleagues via the Internet.

THE INTERNET AND INTERNET-BASED MAPPING TECHNOLOGIES

Web servers and service-based Internet map servers now permit the publishing of cable-related data in high-quality maps on the Internet. These technologies thus enable the distribution of up-to-the-minute RPL data to restricted groups, semirestricted groups, or the general public. This capability can be used for a wide variety of purposes, including permitting programs, collaborative route-selection between project partners, route review by technical experts, or for cable-awareness purposes.

RELATIONAL DATABASE MANAGEMENT SYSTEMS (RDBMS)

Commercial RDBMS software is at the heart of any modern data management strategy. It is a mature means of handling large volumes of complex data in many different industries, such as inventory control, insurance, banking, and transaction processing. Commercial-scale databases bring a level of formality, control, and security to data management that is not possible with a file-based system, regardless of how organized that system may be.

RPL data and submarine cable data in general is inherently spatial, and the spatial portion of the database is indexed to two values: latitude and longitude. This problem—storing spatial data in relational databases—is particularly relevant to databases that must allow access at interactive speeds. One solution is to use multidimensional indexing technology.

CUSTOMIZED SOFTWARE DEVELOPMENT

To synchronize the use of off-the-shelf GIS, RDBMS, and Web server systems, it is necessary to maintain control over both intersystem communication and the user interface of the overall system. This is accomplished through the use of the C++ and Visual Basic programming languages, as well as several Web-scripting languages. Customized software can therefore be thought of as the glue that holds the overall system in place as a single, coherent technology.

DATA MANAGEMENT AND DISTRIBUTION VIA THE INTERNET

The four technologies outlined above have been combined by Thales GeoSolutions to create an Internet-based submarine cable data management and distribution system (figure 1). At the heart of the system is a dedicated cable database built on an off-the-shelf relational database management system. This database constitutes a central repository for all cable data and is available for use in a variety of ways, including the serving of cable data over the Web in cartographic, tabular, or electronic formats. The database also serves as an internal management tool for an organization or consortium, and is fully capable of supporting the dispersed nature of international offices or consortium partners using secure Internet access.

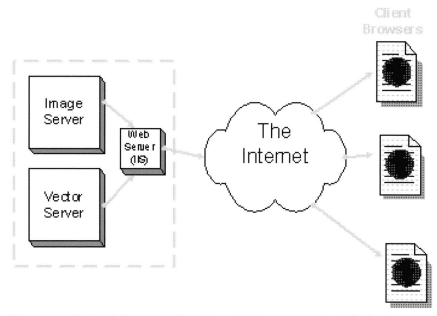

Figure 1. Architectural diagram of Internet-based data management and distribution system.

The associated Internet distribution channels are based on a combination of commercial Web server tools and custom application software that enable specific functionality as required by the customer. The system combines both vector and image data sources in an integrated and seamless environment, and incorporates the latest image compression technology. This means that customers can pan and zoom around large image files, such as sidescan sonar mosaics, at interactive speeds. This image technology is combined with sophisticated streaming of vector data that facilitates interactive client tools. A customer can identify, query, and even edit this vector data from remote sites that access the Thales GeoSolutions server via the Internet.

SECURITY

The principal security concern with a system such as the one proposed here is restricting access of distributed data to authorized users while at the same time allowing public access to specified portions of that data. This security must be substantial due to the value of the data and of the infrastructure described by the data. The first line of defense is standard password/user-name security managed from within the custom-built site. This requires every user to enter a user-name and password in order to be logged onto the system. The usage of the system by each user name will be tracked and any attempts to gain unauthorized entry will be recognized.

Additional levels of security can also be provided depending upon the nature of the data and security requirements of the customer. One such system is provided through the use of Secure Sockets Layer 3.0 (SSL3), which encrypts the packets transferring between client and server, and uses a secure version of HyperText Transfer Protocol (HTTP), known as secure HTTP (or HTTPS), for transport.

Another level of security could be established through the use of encryption certificates. These certificates work by installing an encryption code on a client computer and then checking for this code (or certificate) before allowing access to the system.

CASE STUDY: CABLE PROTECTION

The system described in this chapter is perfectly suited for managing and distributing RPL data via the World Wide Web for the purposes of cable protection. This could be accomplished in a manner that involves at least two separate distribution channels that access the same underlying cable database. The first of these Web-based distribution channels provides cable route data and graphical cable-awareness charts to the public. This would essentially be an "online cable-awareness chart," entirely browser-driven and controlled by the user via a graphical user interface. The second distribution channel would provide technical-level RPL data, possibly including processed engineering data, to authorized customers using a secure Internet connection.

The principal service provided by public-level Web access to the database is the delivery of a graphical cable-awareness chart for a user-selected area, with cable-route data superimposed over bathymetric and other kinds of data (figure 2). The user interface for the public distribution channel presents the user with a global-scale chart from which an area of interest could be selected using simple zoom and pan controls. The user could initiate a query by making successively more detailed graphical selections. Once the user has zoomed in past a specific zoom scale, all of the available cable and background data is displayed on the view. The user may then browse through this information interactively, by panning and zooming, to refine an area of interest. The user may then choose to print the current view to a system printer, effectively creating a cable-awareness chart for the selected area. These self-printed cable-awareness charts (and the display from which they are printed) would be similar in appearance to publicly available cable-awareness charts produced by fisheries and maritime authorities in the United Kingdom and elsewhere. The charts have a professional appearance and include borders, coastlines, bathymetric contours, and other ancillary data, in addition to cable routes. This ancillary data is arranged in thematic layers, which may be enabled or disabled by the user. The user may make the cable-awareness chart as complex or as simple as need be. A demonstration of this concept can be viewed at gis.racal-pelagos.com.

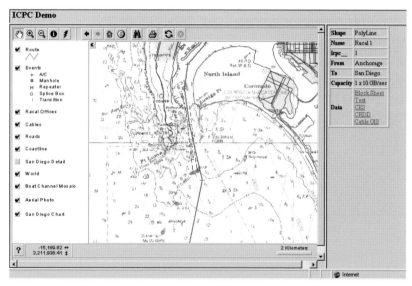

Figure 2. Example of an Internet-delivered "cable-awareness chart."

CONCLUSION   Despite its infancy, the system developed by Thales GeoSolutions and described in this chapter has already been successfully used to provide submarine cable customers with improved data distribution and delivery channels. In general, these technologies and this type of system can be used to provide the following services to the marine cable industry: data distribution and publication; collaborative review and interactive editing; and quality control.

Specifically applied within the submarine cable industry, such a system could be used in the following environments: desktop study; cable route survey, planning, and permitting; and cable installation, maintenance, protection, and repair.

ABOUT THE AUTHORS

Bill Gilmour is the survey manager for the San Diego office of Thales GeoSolutions survey group, responsible for the technical aspects of all marine survey and ocean engineering projects, and the running of the data center. Bill graduated from the University of Glasgow (Scotland) with an honors bachelor of science in topographic science in 1975. He previously has held the office of Secretary of the Hydrographic Society in Scotland, and is an associate of the Royal Institute of Chartered Surveyors. He joined Decca Survey in Great Yarmouth, straight from university, before working in land-surveying activities in the United Arab Emirates. He moved to Aberdeen in 1980 to join Oilfield Hydrographic Projects, which became part of the Stolt Nielsen Seaway Group, as a senior surveyor/programmer, becoming chief surveyor, then area manager for the Indian Office in Bombay. Following three years as a land surveyor with the government of Bermuda, where he ran a pilot digital mapping project and was involved with the introduction of GIS into government, he returned to the Stolt Comex Seaway Group (formerly Stolt Nielsen Seaway Group) as survey manager for worldwide activity. Bill joined Racal Pelagos (now Thales GeoSolutions) as survey manager in Aberdeen in 1996, before transferring in 1999 to the Pacific office of the company in San Diego.

David Millar is vice president of navigation services and commercial software at Thales GeoSolutions (Pacific) in San Diego, California. In this capacity he is responsible for the development of software products and the provision of navigation services to the marine survey and positioning industry. He manages a team of engineers, scientists, technicians, computer programmers, and support staff that primarily supports the submarine telecommunications industry. David graduated from Mount Allison University (Canada) in 1988 with a bachelor of science in math and physics, and received a bachelor of science in survey engineering from the University of New Brunswick (Canada) in 1991. He joined Pelagos Corporation in 1991 as a survey engineer, where he worked mainly on U.S. Navy and commercial cable installation projects. In 1995, he became the navigation operations manager and in 1997, when Pelagos Corporation was acquired by Racal Survey, he was promoted to senior manager of both the software development and navigation operations groups. Over the past ten years, David has been involved in more than 230 submarine cable installation projects and has directed the development of several software products that are now used extensively in the submarine cable installation industry.

CONTACT THE AUTHORS

Bill Gilmour
Bill.Gilmour@thales-geosolutions.com

David Millar
David.Millar@thales-geosolutions.com

Thales GeoSolutions (Pacific)
3738 Ruffin Road
San Diego, CA 92123
Telephone: (858) 292-8922
Fax: (858) 292-5308

Chapter 12

Protected Areas GIS: Bringing GIS to the Desktops of the National Estuarine Research Reserves and National Marine Sanctuaries

DARCEE KILLPACK AND ANDREW HULIN

TECHNOLOGY PLANNING AND MANAGEMENT CORPORATION

NATIONAL OCEANIC AND ATMOSPHERIC ADMINISTRATION COASTAL SERVICES CENTER

CHARLESTON, SOUTH CAROLINA

CINDY FOWLER AND MEGAN TREML

NATIONAL OCEANIC AND ATMOSPHERIC ADMINISTRATION COASTAL SERVICES CENTER

CHARLESTON, SOUTH CAROLINA

ABSTRACT ▶ The Protected Areas Geographic Information System (PAGIS) is a National Oceanic and Atmospheric Administration (NOAA) National Ocean Service (NOS) project designed to assist National Estuarine Research Reserve and National Marine Sanctuary managers in developing spatial data capacity and expertise. The goal of the project is to more firmly support management and research decisions involving protected areas. The NOS PAGIS team, lead by the NOAA Coastal Services Center, is providing each reserve and sanctuary with a fully integrated GIS. The team is compiling GIS data layers for each protected area and providing technical assistance and training in ArcView GIS, ArcView Spatial Analyst, and Global Positioning System (GPS) technology to reserve and sanctuary staff. The project team has also been developing an Internet GIS mapping application, as well as specialized support tools, to address management issues specific to each sanctuary and reserve.

INTRODUCTION

The United States National Estuarine Research Reserves System and the National Marine Sanctuary Program are dedicated to better understanding and protection of the nation's coastal habitats. The Protected Areas Geographic Information System (PAGIS) project (figure 1) is the result of a multiyear initiative to bring GIS technology to bear on research, educational efforts, and general questions or problems of management within these pristine reserves and sanctuaries. The National Oceanic and Atmospheric Administration (NOAA) Coastal Services Center is taking the lead technical role, in conjunction with NOAA's Estuarine Research Division, National Marine Sanctuary Program, and Special Projects Office, in developing this nationwide GIS capacity.

Protected Areas GIS

Figure 1. The PAGIS logo.

The National Estuarine Research Reserve (NERR) system, created with the 1972 passage of the Coastal Zone Management Act, is composed of twenty-five individual sites throughout the United States and Puerto Rico (figure 2). These sites provide opportunities for long-term monitoring and research of estuarine habitat and natural resources, and each is usually involved in many different projects at once, ranging from studies on the natural and human-induced change in the ecology of estuarine ecosystems, to establishing local networks of continuous water-quality monitoring stations. Through the NERR system, a nationwide database of baseline environmental conditions at each reserve has been developed. And through the development of a standardized GIS, these databases are more widely available and consistent across the reserves.

National Estuarine Research Reserves

Figure 2. Map showing the locations of the twenty-five National Estuarine Research Reserves throughout the United States and in Puerto Rico.

The National Marine Sanctuary (NMS) program manages the stewardship of the nation's thirteen protected marine areas through education and outreach programs, resource protection and enforcement initiatives, and science-based conservation management. It also works to conduct and support scientific research and monitoring of the diverse marine and cultural resources within the protected areas (figure 3). One sanctuary, for instance, may concentrate on protecting the breeding grounds of humpback whales, while another focuses on the remains of historical shipwrecks. All of the NMS's projects lend themselves especially well to applications of GIS technology.

National Marine Sanctuaries

Figure 3. Map showing the locations of the thirteen National Marine Sanctuaries throughout the United States and in American Samoa.

PROJECT GOALS

The results of a survey of the twenty-five reserves and thirteen sanctuaries indicated that many kinds and levels of GIS knowledge and expertise existed among and within the various groups and staff. One of the main goals of the project was to provide a basic infrastructure and level of expertise at all of the protected areas. The fundamental components of any spatial data system—hardware, software, data, and applications—all needed to be addressed, and, perhaps most importantly, staff needed to be trained. Project designers consequently identified the following steps as primary: procuring GIS hardware and software; providing GIS, metadata, and GPS training; developing a framework for spatial data layers; creating and maintaining a Web page and listserver; creating custom GIS applications; and providing technical support to all the protected areas.

HARDWARE AND SOFTWARE

Sufficient funds were gathered to buy fully equipped, durable, GIS-friendly computer systems with Internet connections for each reserve and sanctuary. ESRI software, including ArcView GIS, ArcView Spatial Analyst, and ArcPress™, was provided, along with large-format desktop printers. Headquarters of the NERR system and Sanctuary Program were supplied with the same GIS hardware and software packages as well, ensuring a systemwide view of the protected areas.

In addition to the basic GIS software and hardware, GPS receivers were found for those reserves lacking them. The PAGIS team worked with the reserves to develop GPS applications, and provided a customized GPS training manual for them to use. The manual detailed the theory of GPS, data collection techniques, data-processing steps, and integration of GPS data within a GIS. The PAGIS team also provided hands-on GPS training to the reserves through site visits.

TRAINED STAFF

Well-trained staff are crucial to any GIS project. To ensure each reserve and sanctuary could make good use of the hardware and software, formal GIS training was provided to permanent staff members at each site. ESRI-certified instructors at the Coastal Services Center conducted training in basic and advanced ArcView GIS.

Federal Geographic Data Committee (FGDC) metadata training also was provided to assist the reserves and sanctuaries in documenting their data accurately. Finally, a training module was developed specifically to complement the needs of educators within the NERR system. This training session focused on using GIS for educational and presentation purposes, primarily to create and disseminate interpretive maps. Overall, the PAGIS project has provided ArcView GIS, metadata, and GPS training to more than fifty reserve and sanctuary staff. Basic and advanced training continue to be provided as sites grow and change.

TECHNICAL SUPPORT

To handle glitches and daily surprises, reserve and sanctuary staff needed ongoing technical support. At many sites, it is common for only one person to be using the GIS. Access to outside GIS expertise proved essential for the successful development and implementation of the system. Additional technical support was provided to all sites via a Web page, a listserver, telephone and e-mail support, and site visits.

The PAGIS Web site (figure 4) provides, along with a description and status report of the project, an interactive GIS application for viewing spatial data sets, a "GIS in Action" section, an archive of the submissions to the PAGIS listserver, and links to data and GIS resources.

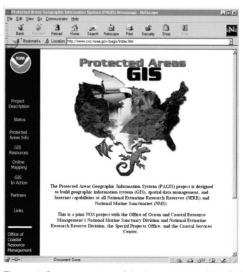

Figure 4. Screen capture of the home page of the PAGIS Web site at www.csc.noaa.gov/pagis.

The PAGIS Internet mapping application couples MapObjects software from ESRI with Microsoft Visual Basic, allowing users to view, pan, zoom, and query different data layers for different sites (figure 5). The mapping application serves as an example of how compatible GIS and the Internet are. It also allows many different users to see the wealth of data sets available to site managers.

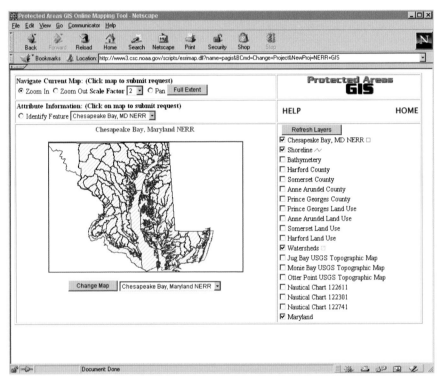

Figure 5. Screen capture of the PAGIS online mapping tool, showing data layers available for the Chesapeake Bay National Estuarine Research Reserve in Maryland.

The "GIS in Action" section of the Web site showcases GIS activities within the reserves and sanctuaries. It provides descriptions of reserve or sanctuary activities, the data sets needed or created, and how GIS and GPS were used. Some examples include:

- the Delaware NERR using GIS to create maps for its site characterization

- the Sapelo Island NERR in Georgia mapping oyster beds as a long-term biological indicator of estuarine environmental health

- the Apalachicola Bay NERR in Florida mapping benthic habitats for monitoring activities, future change assessments, and resource management planning
- the Channel Island NMS using GIS to site possible marine reserves within its boundaries.

The PAGIS listserver is an e-mail forum where GIS and spatial data-related questions can be asked. Most of the questions asked relate to software or hardware issues, but data resources and notices of events are also posted. The listserver also works as a forum for general GIS networking. The listserver is monitored by the Coastal Services Center technical staff. The message archives and the explanation about how to join the list may be obtained from the PAGIS Web site.

In addition to Internet and telephone support, the PAGIS technical support team traveled to reserve and sanctuary sites to lend specific kinds of help, and a "doctor's clinic" was provided at two of the reserve's annual meetings. The clinic provided reserve personnel the opportunity to discuss project-related questions or to ask an expert general GIS and spatial data questions.

Technical support and site visits were provided as well under the umbrella of the joint NOAA and National Geographic Society's Sustainable Seas Expedition (SSE) project. SSE is an initiative to explore and study the marine ecosystems and natural resources within NMS sites. Technical staff were on hand to provide GIS support for data development and on-site training whenever SSE visited a sanctuary.

DATA DEVELOPMENT

Another fundamental and time-consuming step in a GIS project is the development and maintenance of good data. In an effort to build self-sufficiency, technical staff would put the reserve or sanctuary in contact with appropriate state or federal agencies that could supply spatial data and other resources directly to them. If no partnership could be formed, the PAGIS GIS staff compiled and distributed the data sets to the reserve or sanctuary.

Based on the initial assessment and data search, an accurate digital boundary data layer for both the reserves and sanctuaries was considered a high priority. Many of the reserves, however, did not have any digital boundary data, while others had only outdated versions.

Many sanctuaries had digital boundaries that were poorly documented and which frequently differed from the legal boundary descriptions of the sanctuaries as described in the *United States Code of Federal Regulations* (CFR; U.S. National Archives and Records Administration, 2000). Updated and accurate digital boundaries for both the reserves and sanctuaries became an early focus for the team.

The process for reserve boundaries was completed in three steps. The first involved sending each reserve the best available draft version of its boundary to be edited or changed. They were printed on U.S. Geological Survey (USGS) 7.5-minute topographic maps for ease of use. Reserves needing a finer-resolution basemap used USGS digital orthophotoquadrangles (DOQs). If no digital version of the boundary existed, a blank map was sent. Using existing management and research plans as guides, staff edited the boundary by either deleting or drawing new boundary lines on the paper map and documenting those lines.

The paper maps were then sent to the PAGIS team and the process of recording the changes began. To ensure that the lineage of the changes to the digital boundaries could be maintained, a system was designed to attach attributes to each arc: the name of the reserve, the original source of the file, the editor of the file, the status of each arc (i.e., whether it was edited or unchanged), the source of each arc (i.e., the physical ground source by which the arc was moved), the confidence of the edit, and the scale by which each arc was edited. As each arc was changed, either by on-screen digitizing or by using a digitizing tablet, it was coded with these attributes (table 1). Once the edits and changes were verified by the reserve, the final boundary files were compiled and metadata were created.

TABLE 1. AN EXAMPLE OF BOUNDARY COVERAGE, ITS ATTRIBUTE TABLE, AND THE CORRESPONDING CODES FOR THE SOURCE ATTRIBUTE.

Reserve Boundary	Boundary Attribute	TableAttribute Codes	
		Source Code	Source
		10	Road
		11	River/bay edge
		12	Marsh
		33	Railroad
		40	Stream
		42	Trail

As part of the data collection process, other base layers were also compiled for each reserve. These layers included USGS DOQs, USGS digital raster graphics (DRG) files, and NOAA digital nautical charts. These layers, where they were available for a site, and the new digital boundaries, were compiled, placed on a CD–ROM (figure 6), and distributed to each reserve.

Figure 6. Cover graphic of the PAGIS CD–ROM for the Elkhorn Slough National Estuarine Research Reserve, in northern California.

Different obstacles were encountered in the sanctuaries. While numerous sanctuary boundaries had been produced by other parties, none accurately reflected the textual boundary descriptions in the CFR. The PAGIS team and the NMS program decided to develop complete and accurate digital representations of the legal boundaries for all thirteen sanctuaries. The legal boundary of each sanctuary is published in the CFR and is defined with a textual description and a series of coordinates. The objective was to directly translate the listed coordinates and textual descriptions of the boundary into a digital format. Soon after the process was begun, however, it was discovered that many CFR descriptions contained significant errors, ambiguities, and insufficient information to digitally render the boundaries.

Most of the errors arose from conflicts between textual descriptions and listed coordinates in the CFR. For example, the CFR would describe a boundary as following a certain isobath line (i.e., a line representing a single water depth), but when those coordinates were plotted on a NOAA digital nautical chart, they did not coincide with the charted isobath line. Another difficulty centered around the use of adjacent federal and state boundaries as reference points for the sanctuary boundary. In some situations, these other boundaries were used to define the sanctuary boundary itself. Using the other boundaries as references became a problem because many of the boundaries were interpreted from paper maps and were the only source for legal descriptions of the boundaries. If metadata existed for these data sets, the lineage was usually not defined, further blurring the sanctuary boundary. The inshore extent of a sanctuary, and conflicting descriptions of shared sanctuary boundaries, caused problems as well.

The NOS has consequently commenced an extensive analysis of the legal and digital boundary descriptions for all of the sanctuaries. Using ArcView GIS and ArcInfo 8 software from ESRI, boundary coordinate points listed within the CFR were plotted on the largest scale (i.e., highest resolution) nautical chart of the area. Additional data such as park boundaries, marine boundary lines, and digital shorelines was incorporated into the analysis. The described boundaries were then examined in conjunction with adjacent shorelines, bathymetry, and any other features that could have a bearing on the interpretation of the sanctuary boundary.

Once all the issues for each sanctuary are identified, possible resolutions will be developed by the NMS program, including managers and staff from the field sites and headquarters, NOAA General Counsel attorneys, and Coastal Services Center staff. Following agreements among the involved parties, the PAGIS team will begin to create digital boundaries and accompanying metadata.

SPATIAL SUPPORT TOOLS

Now that the reserves and most of the sanctuaries have enhanced GIS capacity, the project has begun to focus on advancing their GIS efficiency with special tools. The initial focus was on creating or finding customized scripts for ArcView GIS that helped reserves and sanctuaries with certain tasks. As GIS skills within the sites advanced, the need for more sophisticated applications developed.

The Channel Islands National Marine Sanctuary, for instance, is currently involved in a joint federal and state process to develop alternatives for siting marine reserves (i.e., restricted access and use areas) within the sanctuary boundary. A working group was formed composed of many different stakeholders, including local resource users, conservationists, and local, state, and federal government staff. The participants are trying to determine the optimal location for a system of marine reserves within the sanctuary boundary, and are using consensus-built criteria and the best available science and socioeconomic information to reach their conclusions. A GIS-based tool was conceived to allow different stakeholders an intuitive visualization and query mechanism to investigate how different reserve locations would affect various resources and areas of recreational and commercial use (Killpack et al. 2001).

The result of the cooperation between the Coastal Services Center and the sanctuary was the Channel Islands–Spatial Support and Analysis Tool (CI-SSAT). CI-SSAT was developed with ArcView 3.2 and Microsoft Visual Basic 6.0 software. CI-SSAT is a framework for compilation and visualization of all the spatial data sets and criteria crucial to siting a marine reserve. It uses a simple suitability algorithm to compute the probability for potential marine reserves, allowing the working group's participants to weight or assign percentages to each criterion based on different perspectives. The group will be able to visualize various management options, weight the quantitative and qualitative benefits and costs of various alternatives, and model the effects of one alternative relative to another. The tool was successfully used in a public forum in September 2000. Plans are underway to transfer the process and tool to other marine protected areas within the NERR system and NMS program that are involved in similar community-focused processes.

CONCLUSION

The PAGIS project is a cooperative effort among various NOAA agencies to bring spatial data technologies to the desktops of the nation's reserves and sanctuaries. This goal was accomplished by procuring GIS hardware and software, providing framework spatial data, creating a support network through a Web site and listserver, conducting site visits, and designing new support tools.

GIS capacity cannot be truly measured by the number of systems installed or staff members trained, but rather on how well the technology is being used day-to-day. This measure of success is harder to quantify, but anecdotal evidence does exist. Increasing interest has developed for specialized and advanced training. For example, the education coordinators in the NERR system have asked for specific training that will enhance their efforts to teach both adults and children. The NMS program headquarters office has also requested additional training for sanctuary staff. Staff from all sites are also posting advanced spatial data and technical questions to the PAGIS listserver, as well as finding other resources of support. They are using the momentum of GIS development to create projects particularly suitable to spatial technologies. They also are capitalizing on many different opportunities to publicize the nature and status of their work outside of the PAGIS project, at conferences with posters and presentations, to other sanctuaries and reserves, as well as state, local, and federal partners. Each of these accomplishments showcases the impact the PAGIS project has had on the NERR system and NMS program, and demonstrates how they have advanced their use of spatial data technologies for coastal resource management in our protected areas.

ACKNOWLEDGMENTS

The authors would like to acknowledge Dwight D. Trueblood from the Cooperative Institute for Coastal and Estuarine Environmental Technology, Mitchell Tartt from the National Marine Sanctuary Program, Ben Waltenberger from the Channel Islands National Marine Sanctuary, and all of the NERR and NMS team members who contributed to the success of this project.

REFERENCES

Killpack, D., B. Waltenberger, and C. Fowler. 2001. The Channel Islands-Spatial Support and Analysis Tool. *Abstracts of the Annual Meeting of the Association of American Geographers.* New York City, Session #3.2.32, CD–ROM.

U.S. National Archives and Records Administration. 2000. *United States Code of Federal Regulations.* Washington, D.C.: Office of the Federal Register, National Archives and Records Administration, United States Government Printing Office.

ABOUT THE AUTHORS

Darcee Killpack has a B.S. in material science from the University of North Carolina–Chapel Hill, and an M.S. in environmental monitoring from the University of Wisconsin–Madison. She has more than six years' experience working with spatial technologies, including GIS, GPS, remote sensing, aerial photography, and custom spatial application programming. She is currently working as a GIS programmer/analyst for the Technology Planning and Management Corporation, as a contractor to the NOAA Coastal Services Center in Charleston, South Carolina.

Andrew Hulin is a coastal management specialist for the Technology Planning and Management Corporation, as a contractor to the NOAA Coastal Services Center in Charleston. He has an M.S. in environmental studies from the University of Charleston, and a B.A. in political science from Drake University.

Cindy Fowler has a B.S. in geography from the University of South Carolina and an M.S. in natural resource information systems from The Ohio State University. She has more than twenty-one years' experience working with GIS, remote sensing, and other forms of spatial technologies. Cindy also has experience in private industry, government service, and university settings supporting the fields of forestry, natural resources, cadastres, geodetic science, and coastal resource management. Currently, she is a senior spatial data analyst at the NOAA Coastal Services Center in Charleston, where she combines her love of the coast with her passion for geospatial technologies. Cindy's research interests are related to coastal and marine GIS, and especially the data needed to support them. She is particularly interested in the technical and legal implications related to marine cadastral data development.

Megan Treml has worked as a GIS analyst at the NOAA Coastal Services Center in Charleston for the past four years. She holds B.S. degrees in biology and geology from the College of Charleston, and is currently pursuing a masters in environmental policy from the University of Charleston and the Medical University of South Carolina.

CONTACT THE AUTHORS

Darcee Killpack
Telephone: (843) 740-1336
Darcee.Killpack@noaa.gov

Andrew Hulin
Telephone: (843) 740-1176
Andrew.Hulin@noaa.gov

Cindy Fowler
Telephone: (843) 740-1249
Cindy.Fowler@noaa.gov

Megan Treml
Telephone: (843) 740-1212
Megan.Treml@noaa.gov

NOAA Coastal Services Center
2234 South Hobson Avenue
Charleston, SC 29405-2413
Fax: (843) 740-1224
www.csc.noaa.gov

Chapter 13

Spatial Policy: Georeferencing the Legal and Statutory Framework for Integrated Regional Ocean Management

ERIC TREML, HAMILTON SMILLIE, AND KIMBERLY COHEN

TECHNOLOGY PLANNING AND MANAGEMENT CORPORATION

NATIONAL OCEANIC AND ATMOSPHERIC ADMINISTRATION COASTAL SERVICES CENTER

CHARLESTON, SOUTH CAROLINA

CINDY FOWLER AND ROB NEELY

NATIONAL OCEANIC AND ATMOSPHERIC ADMINISTRATION COASTAL SERVICES CENTER

CHARLESTON, SOUTH CAROLINA

ABSTRACT In the United States, the policies regulating ocean resources have historically been developed as single-purpose regimes. To facilitate a more integrated, comprehensive ocean management strategy, the National Oceanic and Atmospheric Administration's (NOAA) Coastal Services Center has developed a Web-based regional ocean governance geographic information system (GIS). Working with the four south Atlantic states, the center georeferenced and aggregated ocean policy information, managing agency jurisdictions, and other ocean-use data. This data was then integrated with natural resource data within the system. This paper discusses the issues surrounding the development of these policy and marine boundary data layers in support of regional ocean management and governance. In addition, the Web-based Ocean Planning Information System project, a real-world application of GIS in the marine environment, will be discussed.

INTRODUCTION   Recent history has seen a series of jurisdictional and proprietary claims made on areas of the seafloor, the minerals below the surface, the water column above the seafloor, and the living marine resources contained within the waters. As coastal nations subdivided the ocean and developed governance frameworks, inadequate consideration was given to the spatial integrity of maritime and regulatory boundaries. Today, modern technologies such as GIS, GPS, and the electronic chart display information system (ECDIS), allow ocean resource users to find and exploit marine resources more easily. As resources become more scarce and technologies advance, users will begin to push the envelope of ocean exploitation. Confusing matters further, ocean policies have historically been developed and implemented as single-purpose regimes, with little thought to how they would interact with other resource management considerations (National Research Council 1997). A variety of laws, regulations, and special jurisdictions have evolved over time to protect and manage ocean resources. Unfortunately, this framework is fragmented and complicated, and is often vague with respect to precise geographic boundaries. Consequently, some present-day descriptions of regulatory boundaries are subject to misinterpretation. There is an overlapping horizontal as well as vertical management structure that increases the complexity. This has resulted in a nation ill equipped to address the inevitable conflicts and problems arising offshore.

The ability to visualize regulations, laws, and management structures can assist policy makers in understanding ocean-use conflicts, help to point out inconsistencies in national or state policy, and educate the public on ocean issues. To analyze these ocean policy components, it is essential that geographic information be clearly defined within the legislation. Most of the natural resource data is available in map form, but U.S. laws and regulations have not generally contained the necessary components to map their spatial extents adequately.

To help establish functional digital boundaries that can be used to improve ocean management strategies, the National Oceanic and Atmospheric Administration (NOAA) Coastal Services Center, working with four states in the southeastern United States (figure 1), has been examining existing boundaries and their spatial accuracy. This is part of an ongoing project to develop a Web-based regional Ocean Planning Information System (OPIS) in support of a more coordinated, regional approach to ocean planning and governance. Integral to the functionality of OPIS is the inclusion of georeferenced regulatory data in conjunction with natural resource data. This has involved researching the federal and state policy framework in the region and creating spatial footprints of the geographic area where individual policies apply. These policies and their footprints of geographic applicability are referred to as georegulations. Because policy makers, not geographers, generally write regulations, the challenge is to capture the full and accurate scope of the geography concerned in each individual law. These georegulations, once created, allow users to visualize the spatial extent of regulatory and offshore boundaries and see clearly where natural resources data overlaps them. Using GIS, marine resource managers can identify spatial relationships and see gaps and overlaps in marine policy.

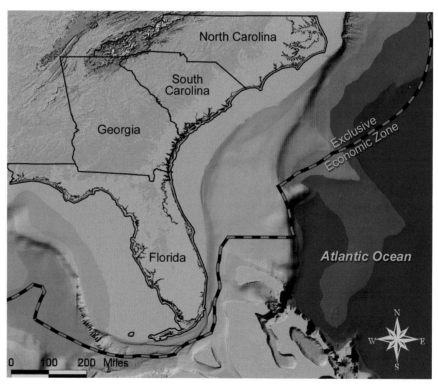

Figure 1. The Ocean Planning Information System (OPIS) was developed for the southeastern U.S. states of North Carolina, South Carolina, Georgia, and Florida. The study area extends from the landward boundary of each state's coastal zone to the outer limit of the U.S. Exclusive Economic Zone.

THE OCEAN POLICY FRAMEWORK

The linking of policy with geography for OPIS required a technical and geographic analysis of the marine boundaries, or cadastre, and the regulatory structure that applies to the area. The term cadastre has not often been used in the context of the marine environment, though many (and some may argue all) of the cadastral components, such as adjudication, survey, and owner rights, have a parallel condition in the ocean. A marine cadastre, similar to its land equivalent, describes the geographic extent of current property interests, as well as past, current, and future rights and interests in the ocean. This includes the delineation of private, state, tribal, national, and international rights. Unlike the land-based counterparts, however, marine geography does not lend itself to simple demarcation with benchmarks, stakes, hedgerows, or fences.

Generally under U.S. law, upland private ownership extends to the mean high-water line. Seaward of this line are the public trust lands that are managed by the state or federal government. In most cases, the individual coastal state has control of the sea bottom and marine resources from mean high water out to the state's seaward boundary (i.e., the Submerged Lands Act boundary) at 3 nautical miles from the coastline (43 U.S.C. §1312, unless reserved by the federal government, §1313). As described in the United Nations Convention on the Law of the Sea (United Nations 1983), the federal government has sovereign rights over all living and nonliving resources out to the extent of the continental shelf and/or exclusive economic zone (EEZ). Like many laws, there are exceptions to the rules. In some states and commonwealths (e.g., Texas, Puerto Rico, Gulf coast of Florida), the state has jurisdiction over the natural resources out to 3 marine leagues (9 nautical miles). Within this area, states have the authority to manage, administer, lease, develop, and use the natural resources of the ocean, and the federal government retains control over commerce, navigation, defense, fisheries, and international matters. There are also other special managed areas such as marine protected areas that can overlap both state and federal waters (figure 2).

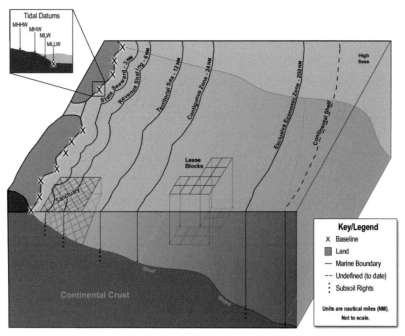

Figure 2. Depiction of U.S. marine boundaries, showing the complexities in offshore jurisdictions. See Fowler and Treml (in press) for an in-depth discussion.

Moving offshore, another layer of shared rights is added with Section 8(g) of the Outer Continental Shelf Lands Act (43 U.S.C. §1337(g)). This act states that 27 percent of all revenues from offshore lease production within 3 nautical miles beyond the state's seaward (i.e., Submerged Lands Act) boundary be given to the states. In many states, the accurate definition of this line has not been an issue, because the exploitation of offshore gas and oil has not been commercially viable. Nevertheless, in some cases, the financial consequences of the location of the boundary can be immense. As an example, in a case often referred to as Dinkum Sands, the federal government and the state of Alaska both claimed rights to offshore revenues (No. 84 Original, U.S. v. Alaska). This case resulted in seventeen years of legal action for the U.S. Supreme Court to establish the right of the federal government to $1.6 billion of oil and gas revenues. Similar Supreme Court litigation has occurred with all but three of the twenty-three U.S. coastal states, the only exceptions being Washington, Oregon, and Hawaii. One can begin to see the complex nature of the multilayered legal framework of state and federal tenure pertaining to the offshore environment and the importance of an accurate marine cadastre.

Beyond a U.S. coastal state's seaward limit (3 or 9 nautical miles), international law and practice take effect, as specified by the UNCLOS. To date, the Unoited States has signed but not ratified the UNCLOS, recognizing the articles and generally abiding by them. In 1988, U.S. President Ronald Reagan signed Presidential Proclamation 5928, which claimed a 12-nautical-mile territorial sea for the United States. As established by the UNCLOS, the territorial sea is a belt of ocean that is measured seaward from the baseline of the coastal nation and which is subject to its sovereignty. This sovereignty also extends to the airspace above and to the seafloor and subsoil. In addition to international law, individual coastal states may exert influence over federal activities within the U.S. territorial sea. Historically, a territorial sea was a 3-mile belt around a sovereign state, reflecting the state of technology of the times (e.g., the distance that a cannon could be shot). A nation can exert exclusive jurisdiction of its territorial sea, except for the rights of innocent passage of foreign vessels.

In addition, under UNCLOS, a nation may claim a contiguous zone, which is an additional 12 nautical miles from the territorial sea. In 1999, U.S. President Clinton extended the U.S. contiguous zone from 12 to 24 nautical miles. This zone allows for an additional area where the sovereign state may enhance law enforcement and public health interests by exercising control over immigration, customs, pollution, and other activities. Extending seaward past the contiguous zone, a nation may claim a fisheries zone for protecting living marine resources, as well as an Exclusive Economic Zone (EEZ) that may extend to 200 nautical miles seaward from the shoreline. The EEZ is an area where a nation may extend protection to both living and nonliving resources above and below the seabed, and assert other associated rights as described under UNCLOS.

The next and relatively uncharted area of the sea that a nation may claim is the area of the continental shelf. Under the UNCLOS, Article 76, countries may claim sovereign rights over the continental shelf more than 200 nautical miles, but to do so must submit technical and scientific details of their claims. The data needed to support this claim is quite complex. The continental shelf boundary determination is one area where the geospatial analyses provided by a cadastral information system may be extremely helpful (Palmer and Pruett 2000). The United States has not made a continental shelf claim to date.

In the United States, federal and state agencies have a variety of overlapping authorities and jurisdictions. Some of the most important federal players include the U.S. Coast Guard (USCG), U.S. Environmental Protection Agency (EPA), U.S. Minerals Management Service (MMS), U.S. Army Corps of Engineers, U.S. Fish and Wildlife Service, and NOAA. This fragmented approach to management of ocean resources at the national level often results in redundant efforts, inefficiency, ineffectiveness, and lack of coordination among agencies with tangled, overlapping jurisdictions. Perhaps more problematic are the unidentified jurisdictional gaps in the existing governance framework that potentially can hinder effective ocean management and appropriate use, conservation, and protection of important ocean resources. These factors emphasize the need for the development of regional spatial frameworks such as OPIS, which will help make sense of offshore jurisdictional complexities (figure 3). To be effective, the content and structure of a digital spatial database must address the important issues of ocean planning and governance, and provide resource managers with a basis for assessing alternative management strategies and tactics.

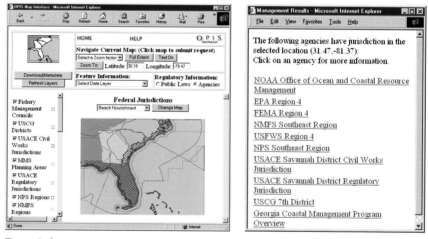

Figure 3. Screen capture of the OPIS Internet mapping application and results window, showing the jurisdictional overlap of many U.S. federal agencies.

**ISSUES ASSOCIATED
WITH CREATING
GEOREGULATIONS**

Clearly there is a need to sort out the policy quagmire, and GIS provides an excellent organizing framework for doing just that. The reality is that very few of the U.S. regulations were written with modern mapping technology in mind. It is not uncommon to find incorrect, imprecise, or inaccurate boundary coordinates published in the U.S. Federal Register and U.S. Code of Federal Regulations. Such instances may be the result of simple human error or misinterpretation of ambiguous legal language, or they may be the result of a lack of understanding of mapping principles. In other cases, the legal description may not adequately describe the geography, or it may be extremely complicated to develop a mapping solution. In a paper cartographic world, the size and scale of the features portrayed on a map or chart mask many spatial inaccuracies and uncertainties. Many times, fundamental cartographic concepts such as scale, resolution, spatial accuracy, datum, and projection are not considered. And because digital mapping technologies enable the development of precise maps, it is essential that those who develop policy understand these important concepts. In that way, the resulting legal framework can be integrated into technologies such as GIS, and the resulting maps made even more accurate.

There are numerous examples of spatial ambiguities in laws and regulations that make mapping difficult. Even when regulations list geographic coordinates (x,y) that should be fairly straightforward, they often fail to list critical information such as the horizontal datum. Many times a boundary will reference a landmark, such as a headland or an ambulatory feature such as the "wash of the waves at high tide." A marine boundary description will often designate a particular isobath (depth contour) or a vertical limit (air space above the ocean). In a digital world, where does one obtain this isobath? Should the line be captured from the nautical chart or should the original sounding data be used to calculate the line? If one does generate this contour or isobath from raw data, what algorithm and parameters should be used? The legal language is not easily interpreted for information system purposes, and requires the user to be both cartographer and detective.

Many of the aforementioned boundaries are drawn some distance from the baseline. The term baseline can be a bit misleading because in actual practice it is not a line but a series of points representing the most seaward portions of the coast (figure 4). Only the most seaward (salient) points affect the location of the offshore boundary.

There are a number of internationally accepted low-water tidal datums from which to determine the baseline, but in the United States the conservative mean lower low-water (MLLW) line is used, as well as closing lines used to separate inland water bodies from the open sea. A number of rule-based spatial determinations are used in establishing closing lines, and help to determine the status of islands and intermittently exposed features fringing the shore (International Hydrographic Organization 1993). The determination of baselines has been, and will most likely remain, a contentious process, as many nations (and many U.S. coastal states) attempt to claim the maximum allowable area. The rules of selecting baseline points and line segments are well beyond the scope of this chapter and the reader is referred to the UNCLOS documents for clarification.

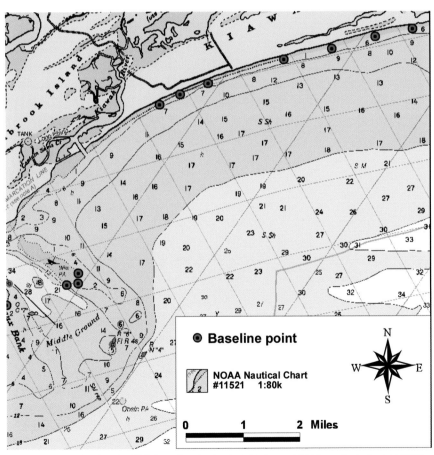

Figure 4. Map showing the U.S. baseline (points) plotted on a NOAA nautical chart (11521) for the area off of Kiawah and Seabrook Islands, South Carolina.

The baseline itself has no real significance to the boundary delimitation issues under consideration here except that it is the basis for the calculation of most offshore boundaries (e.g., territorial sea, EEZ). The calculation for each of the offshore boundaries is mathematically complex. In the paper chart days, a compass was used to swing an arc from each salient point and create an envelope of arcs that could be joined to create the boundary. Today, computers are able to streamline the process and consider each baseline point and its influence on the solution. As the boundary increases in distance from the baseline, fewer baseline points are needed for the solution, and great care must be taken in its development. A simple buffer on a projected map (e.g., Mercator), which is a common operation in most GIS software, is not adequate in this instance and will produce errors. The correct solution must be a three-dimensional projection over the earth's ellipsoidal surface (Ball 1997).

Each point or line segment of a baseline can add enormous significance to the final projected boundary and resulting regulated area. As an example, an island of the smallest significance can lead to a 3-mile boundary covering more than 28 square miles, or an EEZ covering more than 125,000 square miles of adjacent ocean and seafloor (U.S. Department of State 1969; Hodgson 1974). The slight adjustment of a baseline point can result in a large change in the projected area. Another complicating factor is that the baseline, as defined by law, is ambulatory, and therefore most of the offshore boundaries derived from the baseline are ambulatory as well. The exceptions to this are boundaries fixed by Supreme Court decree for Submerged Lands Act purposes. The practical application of describing these ambulatory boundaries for use in a marine information system is problematic to say the least, and often results in confusion for the geospatial data community.

As if the previously described quagmire of jurisdictional boundaries were not complicated enough, there are additional boundaries in the form of special management areas offshore. Examples of these areas include national marine sanctuaries, national parks, and outer continental shelf lease blocks (Thormahlen 1999) to regulate oil, gas, and mineral development. A host of other regulations impart special designations or influence activities in parts of the U.S. oceans.

There are also cases of three-dimensional claims that are difficult to understand, impossible to enforce, and complex to map with the current GIS technology. One of these cases is the NOAA Channel Islands National Marine Sanctuary (water) and the U.S. National Park Service (NPS) Channel Islands National Park (land and water). The NPS has exclusive jurisdiction over the islands and shared administration with NOAA and the State of California's Fish and Game Department from mean high tide out to 1 nautical mile. Within the 1-nautical-mile area, NPS manages the surface waters, while NOAA and the state manage the area below the surface. The state and NOAA share jurisdiction of the entire marine environment from 1 nautical mile to 3 nautical miles (state's seaward boundary), and NOAA has total jurisdiction out to the National Marine Sanctuary (NMS) boundary at 6 nautical miles. Two problems arise: not only is the term in the legal description "mean high tide" not an actual tidal datum and ambulatory, but the inclusion of a vertical (surface waters) jurisdiction requires that the representative cadastral boundary be three-dimensional. This example is difficult to understand and complex to map, and even more complex to manage with regard to the resulting institutional responsibilities.

Other mapping problems can be highlighted as other NMS descriptions are examined. In the case of the Hawaiian Islands Humpback Whale NMS, the legal description uses terms such as "excludes the areas within three nautical miles of the upper reaches of wash of the waves." The Monterey Bay Marine NMS designation states "excludes small areas of cities." Other examples of NMS mapping ambiguities include listing coordinates without reference to a horizontal or vertical datum; referencing natural landmarks; designating a depth isobath as a seaward limit; and citing ambulatory features such as mean high water, river mouths, and the mean low-water line. It should be noted that NOAA, with the assistance of the Minerals Management Service, is working to revise the NMS boundaries, attempting to clear up uncertainty, inconsistencies, and ambiguities that will allow them to be adequately portrayed in a GIS.

Another example is the set of demarcation lines that have been established by the Convention on the International Regulations for Preventing Collisions at Sea, 1972 (commonly called COLREGS). COLREGS define boundaries across harbor mouths and inlets for navigation purposes. If a vessel is landward of the COLREGS line, it must adhere to the Inland Rules of Navigation established under the

Inland Navigation Rules Act. Seaward of the COLREGS line, vessels are subject to rules of navigation established by the International Regulations for Preventing Collisions at Sea, as amended. Historically, the COLREGS lines have been established by drawing a line across a harbor mouth on a paper nautical chart. In addition, geographic references that position these lines are published in the U.S. Federal Register. However, these references typically include no coordinates and often reference ambulatory or ephemeral geographic or man-made features. The georeferencing methods used to describe the location of the COLREGS line are all too often insufficient for use within a digital mapping environment. For example, the following language is used to describe a section of the COLREGS line running from Tybee Island to St. Simons Island, Georgia:

> A line drawn from the southernmost extremity of Savannah Beach on Tybee Island 255° true across Tybee Inlet to the shore of Little Tybee Island south of the entrance to Buck Hammock Creek. (U.S. Department of Transportation, 1995)

From a digital mapping perspective, there are several problems inherent in this approach. For example, beaches are extremely dynamic natural features. Hence, the southernmost extremity of Savannah Beach may shift from storm to storm and year to year. As it does so, a heading of "255° true" may cease to connect the line with the intended location. Furthermore, the reference to the other end of this line segment, "the shore of Little Tybee Island south of the entrance to Buck Hammock Creek," is geographically vague in addition to the fact that it references another ambulatory feature. Finally, no coordinate information is provided that would allow a GIS technician to create a precise and legally accurate map of the COLREGS line.

As these few examples make plain, creating actual digital spatial map layers for use in a GIS is often problematic. In many cases, the legal description of the boundaries may not adequately describe the geography or it may be extremely complicated, requiring subjective interpretation. New legislation and regulations must take into consideration the state of mapping technology. Ideally, the policy regime will list the bounding coordinates and metadata (e.g., datum) with enough precision to create the spatial data layer adequately. Removing the ambiguities is a necessary step in reducing uncertainty for resource analysis such as may be done in GIS applications.

THE OCEAN PLANNING INFORMATION SYSTEM

This effort to outline the boundaries and policy framework is part of an ongoing project to develop an Internet-based ocean governance and management GIS to facilitate the shift in the United States from fragmented management of individual ocean resources to a more integrated, regionwide management approach. This project (OPIS) is available on the Internet at www.csc.noaa.gov/opis. NOAA's Coastal Services Center, working with the four states of North Carolina, South Carolina, Georgia, and Florida, and other federal partners, has been examining existing boundaries, their spatial accuracy, and how these boundaries are used in offshore regulations. OPIS brings this regulatory data together with natural resource data and allows users to view the maps online (figure 5).

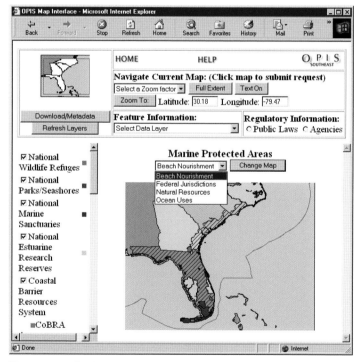

Figure 5. The OPIS Internet mapping application interface allows users to view ocean policy data, jurisdictional boundaries, and environmental data together. This application has four different map views, and includes functions like zoom-in, zoom-out, pan, and identify.

Spatial data within OPIS has been developed using ArcInfo software from ESRI. All regulatory, marine boundary, agency, and environmental data was then integrated using ESRI's ArcView GIS software. The OPIS Internet mapping application was created using ESRI's MapObjects Internet Map Server in conjunction with Visual Basic,

JavaScript™, and HyperText Markup Language (HTML). The entire OPIS Web site consists of more than 350 HTML pages describing recent OPIS project activities, bathymetric collection efforts, beach nourishment issues, U.S. ocean policy, spatial data files, case studies, and project partners. Spatial data files supporting OPIS include dredged material disposal sites, artificial reefs, sand resources, beach nourishment projects by county, NMSs, National Estuary Program sites, data buoys, shorelines, bathymetry, major waterways, outer continental shelf active lease sites, and so on. The geospatial visualization capabilities provided by GIS, in conjunction with the boundary or regulation and natural resource data, creates a uniquely useful tool for the ocean planning and governance community.

The project targets the individual southeastern coastal states and allows the ocean resource manager to examine the significant issues and data of the region in conjunction with the supporting text that describes the laws. The system supports analysis from a particular area of regulated interest, such as marine plastics disposal, or from spatial analysis, such as a particular geographic location (e.g., point on the map). Standard GIS functions are supported, such as view, change layer, pan, zoom, and query. The unique element in this project is the linkage between the policy and its geography. Each applicable data layer contains an attribute link to the appropriate legislation. This allows the user to click on the map area and be presented with the legislation or agency information associated with that particular point (figure 6).

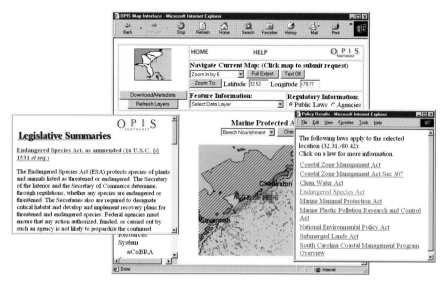

Figure 6. The "Public Law" query, available on the OPIS Internet mapping application, allows the user to retrieve a list of all the ocean-related laws that apply to the area of interest. From this list of links, legal summaries of the legislation are available.

The flexibility of Internet mapping technologies allows the user to drill down to the level of detail needed to satisfy the analysis. For example, the user can look at the attributes of a feature or list the names of the federal regulations related to the point. The user can pick a particular act and look at a synopsis of that legislation or link to the U.S. Code of Federal Regulations for more information. Work is currently underway to convert OPIS to ArcIMS®, which will increase functionality significantly.

CONCLUSION   Spatial analyses of ocean policy can play an important role in balancing the conflicting uses of resources that are occurring in our planet's oceans. Tools such as GIS can help policy makers identify gaps and overlaps in regulations. These types of decision-support tools can lead to better management decisions and more integrated ocean management strategies. In order to conduct the necessary analyses, spatial deficiencies in policy and management regimes must be identified and addressed. New regulations must consider the state of the technology and adequately describe the geography under consideration. Where possible, federal agencies must clear up ambiguities in legal descriptions. The OPIS prototype is an example of what can be accomplished with these goals in place.

ACKNOWLEDGMENTS

The authors would like to acknowledge the support and contributions of the coastal zone management programs of North Carolina, South Carolina, Georgia, and Florida to the success of the OPIS project. In addition, a significant portion of the conceptual and physical design of OPIS was built on prior work conducted by the Florida Marine Research Institute (FMRI). The authors express their thanks to the FMRI staff.

REFERENCES

Ball, W. 1997. *Three Dimensional Coastline Projection Computational Techniques for Determining the Locations of Offshore Boundaries.* Technical Report of the U.S. Department of the Interior, Minerals Management Service, Mapping and Survey Staff.

Fowler, C., and E. Treml. In press, 2001. A marine cadastral information system for the United States. *Computers, Environment and Urban Systems,* Special Issue: Cadastral Systems.

Hodgson, R. 1974. Islands: Normal and special circumstances. In *The Law of the Sea: The Emerging Regime of the Oceans.* J. Gamble, J., and G. Pontecorvo, eds. Cambridge, Massachusetts: Ballinger Publishing Co.

International Hydrographic Organization. 1993. *A Manual on Technical Aspects of the United Nations Convention on the Law of the Sea—1982,* Special Publication No. 51, 3rd Edition, International Hydrographic Bureau. Monaco.

National Research Council. 1997. *Striking a Balance: Strengthening Marine Area Governance and Management.* Washington, D.C.: National Academy Press.

Palmer, P., and L. Pruett. 2000. GIS applications to maritime boundary delimitation. In *Marine and Coastal Geographic Information Systems.* Wright, D. J., and D. J. Bartlett, eds. London: Taylor & Francis.

Thormahlen, L. 1999. *Boundary Development on the Outer Continental Shelf.* U.S. Department of the Interior, Minerals Management Service, Mapping and Boundary Branch. Technical Series Publication, MMS 99-0006.

United Nations. 1983. *The Law of the Sea: The United Nations Convention on the Law of the Sea with Index and Final Act of the Third United Nations Conference on the Law of the Sea.* New York: United Nations.

U.S. Department of State. 1969. *Sovereignty of the Seas.* Geographic Bulletin No. 3, The Office of Strategic and Functional Research, Bureau of Intelligence and Research, U.S. Department of State.

U.S. Department of Transportation, U.S. Coast Guard. 1995. *Navigation Rules, International—Inland,* COMDTINST M16672.2C. Washington, D.C.: U.S. Government Printing Office.

ABOUT THE AUTHORS

Eric Treml has a bachelor of science degree from the University of Wisconsin–Superior in aquatic biology and ecology, and a master of science degree in marine biology from the University of Charleston in South Carolina. He has more than five years' experience working in geographic information science, ranging from coastal marine field surveys and aerial photography interpretation to, most recently, maritime boundaries. For several years, Eric was a GIS spatial analyst/scientist employed by the Technology Planning and Management Corporation, and a contractor to the NOAA Coastal Services Center in Charleston, South Carolina. He recently began doctoral studies in the Nicholas School of the Environment and Earth Sciences, Duke University.

Hamilton Smillie is a spatial analyst for the Technology Planning and Management Corporation, as a contractor to the NOAA Coastal Services Center in Charleston. Drawing on an educational background in geography (bachelor of arts, San Francisco State University, 1991), marine resource management, GIS, and oceanography (master of science, Oregon State University, 1998), Hamilton is involved in projects that deliver remote sensing and GIS technologies to the coastal resource management community. In these products, emphasis is placed on the development of data, tools, and information that will aid state and local officials in coastal resource management decision-making processes.

Kimberly Cohen has a bachelor of science degree in zoology from the University of Rhode Island and a master of arts degree in marine affairs and policy from the University of Miami's Rosentiel School of Marine and Atmospheric Science. She has more than four years' experience working in marine science laboratories and two years' experience with the application of GIS to coastal management issues, such as the identification of offshore sand resources for beach nourishment and the siting of potential aquaculture projects. Currently, Kimberly is a coastal management specialist employed by the Technology Planning and Management Corporation and works as a contractor to the NOAA Coastal Services Center in Charleston.

Cindy Fowler has a bachelor of science degree in geography from the University of South Carolina and a master of science degree in natural resource information systems from The Ohio State University. She has more than twenty-one years' experience working with GIS, remote sensing, and other forms of spatial technologies. Cindy also has experience in private industry, government service, and university settings supporting the fields of forestry, natural resources, cadastres, geodetic science, and coastal resource management. Currently, she is a senior spatial data analyst at the NOAA Coastal Services Center in Charleston, where she combines her love of the coast with her passion for geospatial technologies. Cindy's research interests are related to coastal and marine GIS, and especially the data needed to support them. She is particularly interested in the technical and legal implications related to marine cadastral data development.

Rob Neely is presently serving on a detail from the Coastal Services Center in Charleston to the NOAA National Ocean Service headquarters in Silver Spring, Maryland, where he is developing strategic policy and budget initiatives for NOAA. At the Coastal Services Center, Rob served as project staff member for the Ocean Planning Information System (OPIS). He holds a bachelor of arts in English and economics from the University of North Carolina at Chapel Hill and a master of science in marine resource management from Oregon State University.

CONTACT THE AUTHORS   Eric Treml
Nicholas School of the Environment and Earth Sciences
Duke University
Durham, NC 27708
eat@duke.edu

Hamilton Smillie
Telephone: (843) 740-1192
Hamilton.Smillie@noaa.gov

Kimberly Cohen
Telephone: (843) 740-1181
Kimberly.Cohen@noaa.gov

Cindy Fowler
Telephone: (843) 740-1249
Cindy.Fowler@noaa.gov

Rob Neely
Telephone: (301) 713-3070, extension 191
Robert.Neely@noaa.gov

NOAA Coastal Services Center
2234 South Hobson Avenue
Charleston, SC 29405-2413
Fax: (843) 740-1224
www.csc.noaa.gov

Chapter 14

Geographical Awareness for Modern Travelers: A GIS Application for Maritime Transportation in the Mediterranean Sea

MASSIMO DRAGAN AND MICHELE FERNETTI

UNIDO INTERNATIONAL CENTRE FOR SCIENCE AND HIGH TECHNOLOGY

TRIESTE, ITALY

ABSTRACT Travel has always called for careful planning and a willingness to think on one's feet, but as modern travelers become increasingly demanding in their needs and specialized in their interests, new ways of planning and reacting are clearly being called for. This chapter describes the development of the Ship Information and Management System (SIAMS), which employs ESRI's MapObjects at multimedia information kiosks, as well as ESRI's MapObjects Internet Map Server, to assist international travelers in obtaining real-time ship schedules, retrieving general tourist information on trip destinations, finding connections to other means of transportation, and accessing online booking services (hotels, car rentals, and so forth).

SERVING MARITIME TRANSPORTATION

Maritime transportation plays an important role in the Mediterranean basin, especially so where tourism is concerned. Within that industry, a number of crucial issues—overly long waits for boarding, frequent and often unannounced delays during boarding, lack of organization of port operators, limited communications between authorities and operators, and insufficient integration of maritime data and tourist information systems—are becoming increasingly evident and in need of new solutions. Adequate infrastructures and new telematic services can provide such solutions. Spatially enabled applications can assist in providing timely and accurate data on ongoing and scheduled trips and port destinations. This information can be made available on the Web, at public information kiosks located aboard ships, and at destination/arrival ports, offering the traveler the chance to review, schedule, and plan a journey in real time. The Ship Information and Management System (SIAMS) is an international project within the Telematics Application Programme of the European Commission, the goal of which is to provide these novel services to shipping companies, travel agencies, passengers, citizens, and local authorities through the use of the most advanced technologies in telecommunications, GIS, and networking applications (figure 1).

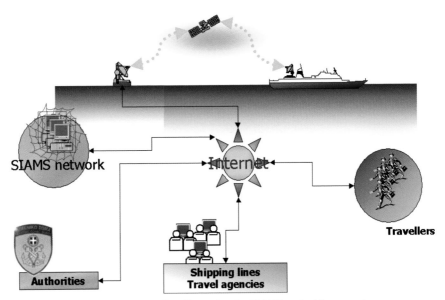

Figure 1. Actors and information flow illustrating the SIAMS project framework.

The MapObjects application makes possible, at any given moment, the provision of information regarding position and condition of ships, and links to other services using spatial data: real-time ship schedules, general tourist information for trip destinations, connections to other means of transportation, and access to online booking services (hotels, car rentals, and so on). GIS-powered services integrate dynamic data and digital maps into SIAMS data services in order to produce real-time trip information and interactive maps of tourist facilities and transportation networks at destination harbors. The services created within this project (figure 2), started in

February 1998, are in use at multimedia kiosks in the ports of Venice and Bari (Italy), Patras, Corfu, Piraeus, Heraklion and Chania (Greece), and onboard a pilot ship (the Aretousa and Minoan cruise ship lines). Through the Internet and a virtual private network, the services are also selectively available to shipping lines, port authorities, travel agencies, and Internet guests.

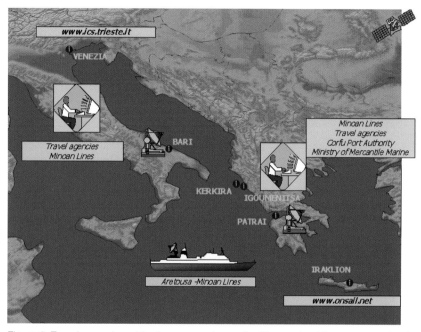

Figure 2. Travel agencies, shipping companies, authorities, and public users were involved in the demonstration phase of the SIAMS project that took place in selected ports in Italy and Greece, and on board the pilot ship.

SYSTEM ARCHITECTURE

SIAMS is built upon the integration of four subsystems that constitute the architecture of the information system. These subsystems entail both hardware (networks and satellite communications) and software (common user interface and GIS interface). The hardware is based on a fully networked infrastructure that connects all the nodes, including the SIAMS server sites and its network, the pilot ship, the multimedia kiosks, the corporate sites (authorities and so on), and Internet access points. Up-to-date networking techniques are employed to provide a secure means to circulate confidential information. Virtual private networks are being used to provide the necessary security level for services with restricted access.

The network makes use of a commercial satellite, Globalstar, and a telecommunications operator, Elsacom, to assure connections with the mobile sites (i.e., the pilot ship). Two earth stations were built, and a gyroscopic satellite antenna with autotracking and stabilization platform was mounted on board an Aretousa ship. Remote control and management tools are included in the system as well.

The development of the third subsystem, the common user interface, dealt with three main areas of concern: application programming, database services, and man–machine interaction. The result is a series of services that require user input and, in the majority of cases, produces a character-based or graphical output. A typical three-tier client/middleware/server architecture using Java™ and Common Object Request Broker Architecture (CORBA®) technology was selected for the sake of multiplatform support. The Java programming language, with all the necessary extensions for communication with the middleware, was chosen for development of the client side. The middleware, based on CORBA, was developed with Inprise™ VisiBroker® 3.2 for Java, while the database system acting as the data repository was designed with the Sybase® database management product.

THE ROLE OF GIS

A GIS environment represents the fourth component of SIAMS. A fully independent, customized GIS application, with typical spatial analysis functionality, powers some of the services. A graphical user interface is also available to the users. The GIS application operates in one of two different modes: as a common GIS customized application, displaying maps with tools and buttons, or in the background, receiving requests from the character-based interface and providing either answers to spatial queries (e.g., estimate the arrival time of a ship) or simple picture maps (e.g., the location of a selected vessel).

In the light of a user requirement analysis carried out at the beginning of the project, it was decided not to choose a fully featured GIS product to fulfill the project tasks, but to opt for a more tailored solution to:

1 integrate the appropriate level of GIS capabilities focusing on the services;

2 provide the necessary level of customization for non-GIS-trained users focusing on the ease of use; and

3 speed up the development, deployment, and performance of the application, focusing mainly on large and detailed data set collections.

Higher efficiency, speed, and reliability can be achieved by creating a brand new application featuring only the needed GIS functionality, instead of customizing an out-of-the-box software package.

ESRI MapObjects software was chosen for several reasons: object-oriented programmability; availability and compatibility of geographical data sets; and the number and quality of compliant GIS software packages (e.g., ArcInfo, ArcView GIS) for the processing and management of geographical data sets. The final reason for choosing MapObjects was the availability of the Internet mapping extension, MapObjects Internet Map Server. Since the common user interface is based on a customized browser application, the map server has been easily integrated into a single environment. This solution allows platform independence and one-step software management and updating or data set modification; it cannot, however, be implemented on board the pilot ship: the network load is too great for the satellite system's bandwidth—at least at this stage. Therefore, two different GIS user interfaces have been set up: a MapObjects stand-alone application on board and at port kiosks, and MapObjects Internet Map Server, which is available to Internet users and to certain authorized users (figure 3).

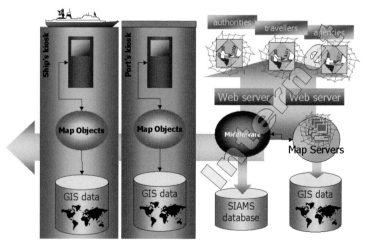

Figure 3. The GIS architecture integrates dynamic data and digital maps into SIAMS services. A stand-alone application is available on multimedia kiosks, and MapObjects Internet Map Server serves Internet clients as well as authorized users.

The Web server and map server are hosted on two separate machines: the former is a Microsoft IIS 4.0 hosted at the International Centre for Science and High Technology (www.ics.trieste.it); the latter is hosted on a Windows NT 4, Pentium® III, 500-MHz workstation with 256 MB of RAM.

BRIDGING DIFFERENT ENVIRONMENTS: GUARANTEEING INTEROPERABILITY

Windows software components are often based on a component object model (COM). The two dominant models are CORBA and Microsoft's object linking and embedding/component object model (OLE/COM). MapObjects is based on the OLE/COM model; other services and the user interface were developed with Java/CORBA. In order to build up an integrated environment, a bridging software (Visual Edge Software Ltd.'s ObjectBridge for COM-CORBA) was included in the system architecture. Three levels of interoperability were built between the GIS interface and the browser-based interface: (1) low-level compatibility, guaranteed by the bridge application; (2) the ability to switch between interfaces; and (3) exchange of parameters and commands based on sockets for interapplication communication (figure 4).

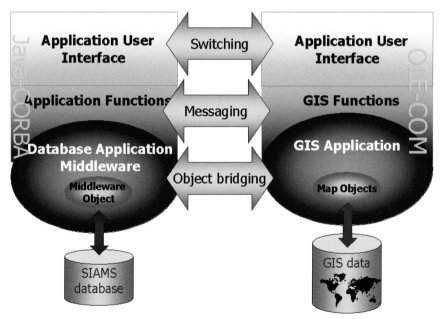

Figure 4. The architectural solution: a CORBA-COM integrated environment communicates at three different levels, guaranteeing component interoperability.

GIS-POWERED SERVICES

SIAMS was designed to cover all the phases of a journey by ship, starting from trip planning at home, through reaching the departure location, boarding procedures, and on-board assistance, to the supply of useful information regarding destination sites. Two categories of services were developed: (1) a set of publicly accessible services available at multimedia kiosks or via the Internet, mainly addressed to travelers for tourism or commercial purposes; and (2) a set of restricted services for shipping lines, travel agencies, port authorities, customs offices, and so forth. Public services provide information related to the trip (ship position, schedules, estimated timetables), booking facilities, tourist information, Internet access on board, announcements, weather conditions, and more. Authorized services include ship management issues (personnel, finance, monitoring), boarding plan administration, real-time trip information (accurate ship positioning, speed, vessel monitoring), and office procedure automation.

Three of these services are supported by MapObjects: ship location (public and authorized), schedule information, and port information. As mentioned earlier, the GIS services have been designed to support two interfaces: a stand-alone interface for the user at a kiosk, and a Web interface available through a browser such as Netscape Navigator® or Microsoft Internet Explorer. The following sections describe these services in detail, along with their intended purposes and the look and feel of their interfaces.

WHERE IS MY SHIP? THE SHIP LOCATION SERVICES

Ship position
One of the basic and most important functions of the system is to represent ship positions to on-board passengers, to users at port kiosks, or to users on the Web (figure 5). Position information retrieved from GPS receivers installed on the vessels is transmitted via satellite and stored in the database where it is accessible to all applications. Security concerns restrict this level of service to display of approximate positions only. The ship's actual coordinates, retrieved from the database, are not shown on the interface, and the zoom level is limited. For the same reason, the level of detail of the background is also limited by the controlled accuracy of the vector and raster data sets used (see geographic representation, figure 5).

Figure 5. The ship location service interface visually delivers different kinds of data to travelers: real-time ship position, arrival time estimation, and weather information. The authorized version also provides confidential information such as exact ship coordinates and velocity.

Authorized ship location

A variant of the above service is the authorized version. Its main purpose is to give a precise representation of a ship's position to shipping companies and port authorities (figure 5). This information is extremely important to ship monitoring and for fast responses to emergencies. The design follows the framework of the public version, but with two extra elements: the actual vessel coordinates for each ship, and vessel speed. Considering the heterogeneous environment in which a stand-alone application could be implemented, the service is available only through the MapObjects Internet Map Server application. The use of the Virtual Private Network is therefore required to guarantee safety and confidentiality.

Arrival-time estimation

Real-time ship locations are used to estimate the arrival times, which are either displayed on the GIS interface or calculated on request and passed on to the user interface. These calculations also make use of vessel passage plans as provided by shipping companies. Passage plans report the route the ship will take, broken down in segments of known length and expected time duration. Conventional GIS spatial analysis procedures are employed to determine the total amount of time needed to arrive at each destination. The GIS interface therefore gives the actual time elapsed since departure, the estimated time left to reach the next port, any subsequent stops, and a map showing an approximate location of the ship. Real-time weather information along the route is also displayed.

Geographic representation

The geographical background on which the ships are represented is a combination of a raster digital terrain model (DTM) overlain with thematic vector data derived from the Digital Chart of the World data set (1:1,000,000 scale). The DTM data is available from the National Geophysical Data Center Web site and has a spatial resolution of 30 arc seconds. ArcInfo was used to mosaic the different tiles that cover the Mediterranean Sea region, and to integrate vector data depicting country and province boundaries, major cities, airports, sites of interest, rivers, and so on. The public version of the location service is implemented on both the kiosk stand-alone platform and with MapObjects Internet Map Server.

WHEN IS MY SHIP CRUISING? THE SCHEDULING SERVICE

The SIAMS database is structured to manage, analyze, and retrieve information regarding all schedules for travel by ship in the Mediterranean basin. A Java-based user interface allows users to retrieve information on all available connections between a departure and a destination port with a specified date and time lag (figure 6). A multistep window menu leads to a list of all ships covering the specified itinerary with a detailed timetable showing departure and arrival times, trip duration, and intermediate stops. Optional links allow the user to retrieve GIS-supplied information. For this service, the GIS application provides a map of the ship's entire itinerary highlighting the user's specified section, the current position of the ship, and the official timetable. The service is available both at the kiosk stand-alone platform and within MapObjects Internet Map Server. A visually driven selection of itineraries to obtain scheduling information is currently under development.

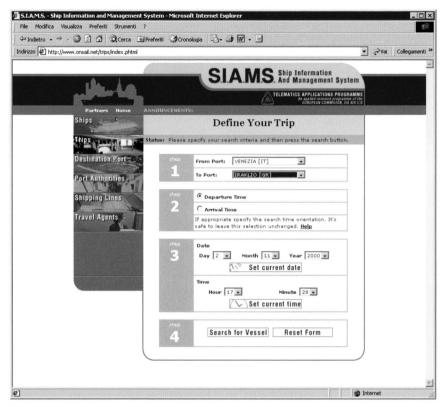

Figure 6. The schedule service interface allows passengers to plan their trips or find information about cruising ships, along with maps that show itineraries or selected ship locations.

WHAT WILL I FIND THERE? THE PORT INFORMATION SERVICE

GIS is also used to offer travelers sets of useful information on the ports and surroundings for points of departure and arrival. The GIS interface provides a detailed map of a selected port area, directions to reach a port or other destinations, road networks and connections to other means of transportation, sites of interest, weather information, and so forth. Navigation tools for visualization at multiple levels of detail (district, town, street) and search capabilities (attribute-based and spatial queries) support the user. Hot links to multimedia information (pictures, text, and so on), additional Web sites (hotels, museums, and so on), and other tourist information systems for online booking are also supplied.

A very critical issue for this service has been the retrieval of highly detailed digital data sets of the ports and surrounding areas of the sites involved in the pilot project (Venice and Bari for Italy, and Corfu, Igoumenitsa, Patra, and Chania for Greece). Such data is crucial to the creation of maps at an adequate scale and to the localization and display of facilities of tourist interest. TeleAtlas provided the data for the Italian sites (1:10,000 scale), while the ESRI Greek distributor supplied the data on Greek ports.

WHERE DOES GIS MAKE THE DIFFERENCE?

Transportation information systems such as SIAMS are inherently geographical, making GIS technology critical to their success. A number of features in this regard, peculiar to SIAMS, are worth highlighting (figure 7):

- a friendly (geo)graphical user interface to explore the information space;

- the integration of dynamic data and digital maps, which makes it possible, for example, to provide real-time information on distances and delays;

- interactive mapping available on request to both kiosk and Web users;

- provision of spatial queries and analyses based upon the relationships between entities (e.g., route estimations, finding features within a certain radius) are typical GIS tasks that can easily be performed; and

- background spatial support to the common user interface, provided as an alternative operative mode.

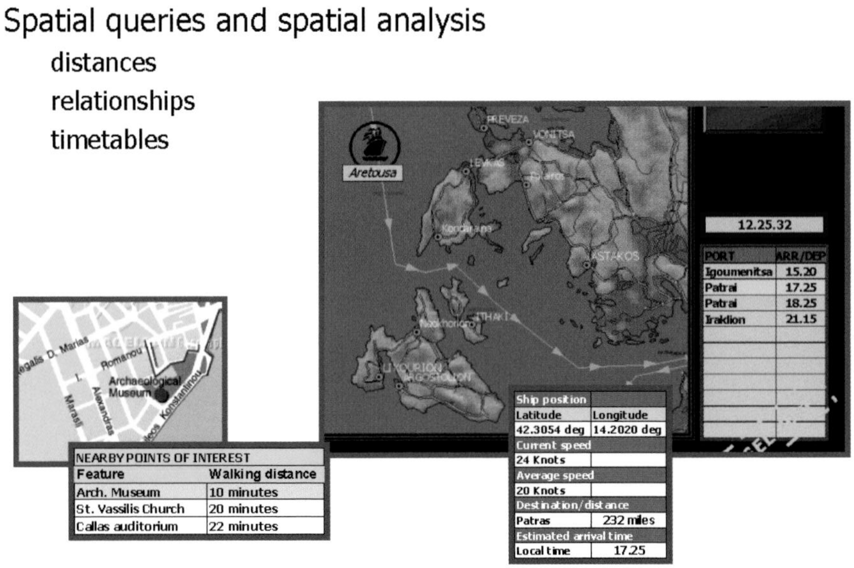
Spatial queries and spatial analysis
distances
relationships
timetables
Aretousa
PREVEZA
ASTAKOS
ITHAKI
12.25.32
PORT ARR/DEP
Igoumenitsa 15.20
Patrai 17.25
Patrai 18.25
Iraklion 21.15
Ship position
Latitude Longitude
42.3054 deg 14.2020 deg
Current speed
24 Knots
Average speed
20 Knots
Destination/distance
Patras 232 miles
Estimated arrival time
Local time 17.25
Archaeological Museum
NEARBY POINTS OF INTEREST
Feature Walking distance
Arch. Museum 10 minutes
St. Vassilis Church 20 minutes
Callas auditorium 22 minutes

Figure 7. GIS-powered services support the delivery of information through the (geo)graphical user interface by means of spatial queries and analysis.

FUTURE DEVELOPMENT

The setting up of an exploitation plan is one of the primary goals of the project. To that end the architecture of the system has been structured to be open and scalable, capable of catering to multiple vessels and multiple shipping lines and providing additional services. Development of GIS components will go in two main directions: toward Web-based services and advanced scheduling capabilities. As cheaper satellite connections and faster Internet connections become available, Web services will be based entirely on Internet mapping technology. Improvements in intermodal transportation scheduling will take advantage of the network analysis algorithm components in ESRI NetEngine™ software, to perform tasks such as finding the best or the cheapest route, combining multiple means of transportation and multiple shipping companies, and so forth.

ACKNOWLEDGMENTS

The authors thank Enrico Feoli and Gennaro Longo of the Earth, Environment, Marine Science and Technology Area at the International Centre for Science and High Technology–UNIDO. SIAMS is an international project within the Telematics Application Programme of the European Commission. The project is coordinated by Vassilis Spitadakis of FORTHnet SA, Hellenic Telecommunications and Telematics Applications Company, Vassilika Vouton, P.O. Box 2219, GR 710 03 Heraklion, Crete, Greece. Other participants include Alfa Ltd. (Greece), Visual Software Italy srl (Italy), TOP-REL srl (Italy), and the Instituto Balear de Innovacion Telematica (Spain). The International Centre for Science and High Technology (ICS–UNIDO, Italy), is responsible for GIS services implementation. Associated partners are Minoan Lines Shipping SA (Greece), the Greek Ministry of Mercantile Marine, High Technologies Associate Ltd. (United Kingdom), and Systron srl (Italy). Sponsoring partners for Tourist Information Systems are Regione Puglia (Italy), Egnatia Foundation and Region of Epirus (Greece), Region of Crete (Greece), and the provinces of Drama, Kavala, and Xanthi (Greece). Sponsoring partners for GIS software and geographic data are ESRI Italia, TeleAtlas (Italy), and Marathon Data Systems (Greece).

PROJECT WEB SITES

ESRI ArcIMS: www.esri.com/software/arcims (now incorporates MapObjects Internet Map Server)

ESRI MapObjects: www.esri.com/software/mapobjects

International Centre for Science and High Technology Internet Map Server site: www.ics.trieste.it

SIAMS project Web site: www.onsail.net

ABOUT THE AUTHORS

Massimo Dragan, Ph.D., is a researcher at the department of biology, University of Trieste. He holds a bachelor of science degree in natural sciences, as well as a doctoral degree in biomonitoring methods and environmental quality. Massimo lectures in environmental information systems, and his research interests are in the fields of landscape ecology, wildlife management, and environmental data analysis. He currently works on national and international projects dealing with GIS and remote sensing applications for natural resources conservation.

Michele Fernetti is responsible for all technical operations in the laboratory of geomatics and GIS at the University of Trieste. He has a bachelor of science degree in natural sciences and is currently working toward a Ph.D. in biomonitoring methods and environmental quality. Michele also works as a consultant in national and international projects dealing mainly with natural resources management and related spatial information systems. He lectured for two years in automated and thematic cartography, and his main research activities focus on digital cartography, GIS, and remote sensing for environmental assessment.

CONTACT THE AUTHORS

Massimo Dragan, GIS consultant
dragan@univ.trieste.it

Michele Fernetti, GIS consultant
fernetti@univ.trieste.it

UNIDO International Centre for Science and High Technology
GEOLAB - University of Trieste
c/o CSIA- Edif. H2
Via Valerio 12,
34127 Trieste, Italy
Telephone: +39-040-676-3304
Fax: +39-040-676-3316

Selected Bibliography and Internet Resources

The "undersea world of GIS" currently enjoys a constant stream of publications that follow the latest developments in this dynamic new field. Some of the earliest papers (late 1980s, early 1990s) were published almost exclusively in oceanographic or earth science journals, but recent works appear more frequently in specialty GIS journals. The first complete book about ocean GIS was published by Taylor & Francis in late 1999, with a second printing appearing in 2000. In addition to perusing the sources below, a good starting point is the "Library" section of the ESRI Virtual Campus, at campus.esri.com. Included there is an online annotated bibliography containing thousands of references to GIS-related journal articles, as well as a "Technical Reports" link to complete texts of hundreds of conference papers and technical reports. And finally, visitors to the ESRI Virtual Campus may enjoy the excellent virtual course by Genevieve Healy, *Integrating Marine Science GIS into a K–12 Classroom.*

BOOKS AND WORKBOOKS

Wright, D. J., and D. J. Bartlett, eds. 2000. *Marine and Coastal Geographical Information Systems.* Research Monographs in GIS Series. London: Taylor & Francis, dusk.geo.orst.edu/book
Launched as a project to mark the United Nations International Year of the Ocean (1998), and supported by the International Geographical Union's Commission on Coastal Systems, this book covers the fundamental issues of representation and data modeling, applications, and institutional issues for marine and coastal GIS, including commentaries on the reliability of data retrieved by various mapping and sampling instruments, and guidelines for avoiding common mistakes in applying GIS to marine and coastal data.

Convis, C. L., Jr., ed. 2001. *Conservation Geography: Case Studies in GIS, Computer Mapping, and Activism*. Redlands, California: ESRI Press.
An impressive compilation of the recent work by scores of nonprofit organizations and conservation groups worldwide who are applying GIS to a host of environmental problems and conservation issues. There is a section entitled *Marine Geography*, compiled and introduced by Joe Breman of ESRI's Marine Conservation Program, and featuring the work of Florida's Dolphin Ecology Project, the International Marinelife Alliance, Canada's Bay of Fundy Resource Centre, the USGS Glacier Bay Field Station in Alaska, the Puget Sound Estuary Habitat Program, the Marine Conservation Biology Institute, Los Angeles County's Stream Team, the Center for Marine Conservation, The Surfrider Foundation, the New England Aquarium, the Oceanic Resource Foundation, Woods Hole Oceanographic Institution, and the FGDC Marine and Coastal Spatial Data Subcommittee.

Earle, S. A., 2001. *The National Geographic Atlas of the Ocean: The Deep Frontier*. Washington, D.C.: National Geographic Society.
National Geographic Society Explorer-in-Residence Dr. Sylvia Earle follows up her highly successful books *Wild Ocean: America's Park Under the Sea* and *Dive: My Adventures in the Deep Frontier* with this offering, the first collection of oceanographic maps, images, and information ever published by the Society. Dr. Earle developed the atlas in cooperation with NOAA, NASA, and the U.S. Navy. Accompanying the book is a geography skill and standards guide for secondary school teachers that includes a section on GIS.

Fisher, W. L., and F. J. Rahel, in press. *Geographic Information Systems in Fisheries*. Bethesda, Maryland: American Fisheries Society (due out late 2002).
This book will discuss challenges of using GIS in aquatic environments and presents applications of GIS for freshwater and marine fisheries.

Lang, L. 1998. *Managing Natural Resources with GIS*. Redlands, California: ESRI Press.
Presents several case studies of real organizations using GIS to address pressing issues in natural resource management, including coastal protection.

St. Martin, K., ed. 1993. *Explorations in Geographic Information Systems Technology, Volume 3: Applications in Coastal Zone Research and Management*. Worcester, Massachusetts: Clark Labs for Cartographic Technology and Analysis, and Geneva, Switzerland: United Nations Institute for Training and Research (UNITAR).
This is an Idrisi/UNITAR workbook with training exercises on coastal applications of GIS.

Valavanis, V. D., in press. *Geographic Information Systems in Oceanography and Fisheries.* London, UK: Taylor & Francis (due out March–April, 2002).

This book will present a summary of GIS concepts applied to fisheries and physical oceanography with major sections focusing on marine GIS applications for cephalopod resources in European seas and the southwest Atlantic. Also included will be programming codes and subroutines for the applications described.

JOURNALS

Although there is still no single journal devoted entirely to ocean GIS, *Marine Geodesy* has perhaps come the closest, having offered three special issues on marine and coastal GIS in 1995, 1997, and 1999 (vol. 18, no. 3; vol. 20, nos. 2–3; and vol. 22, no. 2). *Marine Geodesy* is published by Taylor & Francis, and is searchable at www.tandf.co.uk/journals/tf/01490419.html. Other journals that often cover marine and coastal applications of GIS include:

Computers, Environment & Urban Systems, published by Elsevier, is searchable at:
www.elsevier.com/inca/publications/store/3/0/4

Computers and Geosciences, published by Elsevier, is searchable at:
www.elsevier.nl/locate/cgonline

GEOWorld (formerly *GISWorld*), *GEOEurope,* and *GEOAsia Pacific,* are glossy trade journals published by Adams Business Media, and available at:
www.gw.geoplace.com/gw, www.gw.geoplace.com/ge, and www.gw.geoplace.com/asiapac

Geospatial Solutions (formerly *GeoInfo Systems*) is another glossy trade journal, published by Advanstar Communications, Inc., and available at:
www.geoinfosystems.com

Integrated Coastal Zone Management, published by ICG Publishing Ltd., is searchable at:
www.iczm.org

International Journal of Geographical Information Science, the premiere academic journal in the fields of GIS and GISci and published by Taylor & Francis, is searchable at:
www.tandf.co.uk/journals/tf/13658816.html

Journal of Coastal Research, published by the Coastal Education and Research Foundation, Inc., is searchable at: www.cerf-jcr.com

The Professional Geographer, published by the Association of American Geographers, carried a special issue on "Ocean Space" in 1999 (vol. 51, no. 3). More information on the journal may be found at: www.aag.org

Sea Technology, published by Compass Publications, is a leading glossy trade journal of ocean engineering, design, equipment, and services. It is searchable at: www.sea-technology.com

Surveying and Land Information Systems, published by the American Congress on Surveying and Mapping, carried a special issue on coastal GIS in 1998 (vol. 58, no. 3). It is searchable at: www.acsm.net/publist.html

Transactions in GIS, published by Blackwell Publishers, is searchable at: www.blackwellpublishers.co.uk/journals/tgis

CONFERENCES AND PROCEEDINGS

Happily, there are now several conferences devoted entirely to coastal GIS. The deep ocean (marine) community is much smaller and tends to present research results either at oceanography conferences, or at vendor user conferences with special sessions on ocean research and management.

American Geophysical Union (AGU), Fall or Spring Meetings, held annually in San Francisco and Boston respectively, www.agu.org/meetings. These meetings have had occasional papers or posters involving marine and coastal GIS. The AGU is one of the world's largest scientific societies for specialists in earth, ocean, atmospheric, and planetary sciences. Proceedings from the AGU are published as *Eos, Transactions of the American Geophysical Union.* The society also cosponsors, with the American Society of Limnology and Oceanography, the Ocean Sciences Meeting, held annually at different sites.

Caris GIS User Conferences, New Orleans, Louisiana,
www.caris.com
In the early 1990s, Caris GIS was the first commercial marine GIS package to be
broadly released in North America, and continues to enjoy success, particularly
within the Canadian hydrographic community. Proceedings of these conferences are
available on CD–ROM from Caris.

Coastal GeoTools, NOAA Coastal Services Center, Charleston,
South Carolina, www.csc.noaa.gov/GeoTools
The Coastal GeoTools conference series began in 1999 and will be held biennially in
Charleston. The event is designed to help coastal resource managers make better use
of spatial technology, particularly GIS, the Internet, remote sensing imagery, metadata, and GPS. Proceedings of these conferences are available at the Web site above
or on CD–ROM.

Coastal Zone Canada, held at different sites annually by the Coastal
Zone Canada Association, www.dal.ca/aczisc/czca-azcc

Coastal Zone (U.S.), held at different sites biennially,
www.csc.noaa.gov/cz2001
This event is now the largest meeting of coastal resource managers in the world.

CoastGIS, an international symposium held at different sites in various years, www.coastgis.org
Halifax, Canada, was the site for the 2001 symposium, which built on the successes
of previous gatherings in Cork, Ireland; Aberdeen, Scotland; and Brest, France.

ESRI User Conference: Coastal, Ocean, and Marine Resources
Track, held annually in San Diego, California, www.esri.com

PACON (PAcific CONgress on Marine Science and Technology),
held at different sites annually, www.hawaii.edu/pacon

PACON 2001 was held in San Francisco and featured a session on
marine and coastal GIS within the "Ocean Science and Technology"
track.

OCEANS, held at different sites annually by the IEEE Oceanic Engineering Society and the Marine Technology Society, www.mtsociety.org/
conferences/index.cfm.

WEB SITES

Many of the preceding references include accompanying Web sites, but the following sites provide a wealth of additional resources in terms of data, publications, case studies, and tools:

Annotated Bibliography of Coastal GIS (updated every two years), www3.csc.noaa.gov/gisprojects/biblio

ArcGIS Marine Data Model
dusk.geo.orst.edu/djl/arcgis

Davey Jones Locker
dusk.geo.orst.edu/djl/links.html
A fairly complete listing of Web sites focused on marine/coastal GIS, as well as sea-floor (seabed) mapping and visualization, updated almost daily. At the time this book went to press, there were more than seventy links to marine/coastal GIS resources and nearly eighty to seafloor mapping sites.

ESRI Marine Conservation Page
www.esri.com/conservation/links/marine1.html
Seven pages of useful, annotated links to sites of interest for ocean mapping/GIS, ocean GIS scholarly papers and ESRI conference proceedings, and public ocean conservation and GIS data for download.

Florida Marine Research Institute (FMRI) GIS and Mapping
www.floridamarine.org/features/category_main.asp?id=1153
FMRI was selected from more than sixty thousand organizations worldwide to receive a 2000 Special Achievement Award in GIS from ESRI, for its effective use of GIS in marine conservation.

FMRI's Statewide Ocean Resource Inventory (SORI)
ocean.fmri.usf.edu/ims/sori

Oceansp@ce
www.oceanspace.net
This site is now among the world's largest online newsletters of marine science and ocean technology, including many job listings and announcements pertinent to marine/coastal GIS.

Nautical Data International, Inc.
www.ndi.nf.ca
The company specializes in the production and distribution of digital hydrographic and other data products to serve the needs of GIS users.

NOAA NOS Office of Coast Survey, Electronic Navigational Charts
chartmaker.ncd.noaa.gov/ocs/enc/vector1.htm

NOAA Pacific Marine Environmental Laboratory Visualizations Page
www.pmel.noaa.gov/visualization

SEA-GIS Listserv, one of the world's largest listservers for discussion of marine and coastal GIS issues, data, employment, and so forth.
listserv.heanet.ie/sea-gis.html

Sylvia Earle at the 1999 ESRI Special Ocean Exhibition
www.esri.com/news/arcuser/1099/deepness.html

SOME OCEAN GIS DATA SITES

The URLs for these sites may change in the future, but one may always do a search for the titles below on any major Internet search engine, such as Yahoo, AltaVista, Google, and so on.

Central Coast Joint Data Committee (California)
www.centralcoastdata.org

Cook Inlet Information Management & Monitoring System (Alaska)
info.dec.state.ak.us/ciimms

CoastBase, "The European Virtual Coastal and Marine Data Warehouse"
www.coastbase.org

Delaware Spatial Data Clearinghouse
gis.smith.udel.edu/fgdc2/clearinghouse

Dynamic Estuary Management Information System (DEMIS, Oregon)
www.lcd.state.or.us/coast/demis/core.htm

Fagatele Bay National Marine Sanctuary (American Samoa) GIS Data
dusk.geo.orst.edu/djl/samoa

Federal Geographic Data Committee (FGDC) National Geospatial Data Clearinghouse
130.11.52.184

Geography Network
www.geographynetwork.com

Gulf of Mexico GIS and Internet Map Server (USGS)
sheba.er.usgs.gov/gulf

InfoRain (Pacific coast data from Ecotrust)
www.inforain.org

National Geophysical Data Center's NOAA Server
www.ngdc.noaa.gov/NOAAServer

NOAA Coastal Services Center
www3.csc.noaa.gov/CSCweb/tempProdCat.asp

NOAA Coastal Services Center Clearinghouse
clearinghouse.csc.noaa.gov/NOAA_CSCgateway.html

NOAA CoastWatch Satellite Data
cwatchwc.ucsd.edu/data.html

NOAA Nautical Charts
anchor.ncd.noaa.gov/noaa/noaa.html

NOAA NOS Mapfinder
mapfinder.nos.noaa.gov

NOAA's Ocean Resources Conservation & Assessment (ORCA)
cammp.nos.noaa.gov/spo/prodlist.taf?alltype=3

NOAA Pacific Marine Environmental Lab, Vents Program (data
from the Juan de Fuca Ridge, 300 miles off the Oregon/Washington
coast)
newport.pmel.noaa.gov/gis/data.html

Ocean Planning Information System (OPIS, southeast United States)
www.csc.noaa.gov/opis

Olympic Peninsula Clearinghouse (Washington)
cathedral.cfr.washington.edu/~chouse

Oregon Coast Geospatial Clearinghouse (Oregon)
buccaneer.geo.orst.edu

Protected Areas GIS (PAGIS, National Marine Sanctuaries and
National Estuarine Research Reserves)
www.csc.noaa.gov/pagis

Statewide Ocean Resource Inventory (Florida Marine Research Institute)
ocean.fmri.usf.edu/ims/sori

Teale Geographic Library (California)
www.gislab.teale.ca.gov/wwwgis /dataview.html

USGS EarthExplorer
earthexplorer.usgs.gov

USGS National Mapping Geospatial Data Clearinghouse
mapping.usgs.gov/nsdi

Wisconsin Coastal Clearinghouse
www.lic.wisc.edu/coastgis

Undersea with GIS
Book design by Michael Hyatt
Book production and image editing by Jennifer Johnston
Cover by Amaree Israngkura
Printing coordination by Cliff Crabbe